READ AT A GLANCE

All Learning is Self-Teaching

How do you or does everyone 12-90 years old in the family relate to the following?

1. (a) Do you sometimes write checks in words and numerals for your bank account?.
   (b) Do you know how to accomplish (a)?
   If No, you need to study arithmetic.

2. (a) Do you sometimes change measurements in yards to meters and vice versa?
   (b) Do you know how to accomplish (a)?
   If No, you need to study arithmetic.

3. (a) Do you sometimes have to determine for example, the elapsed time from 1:30 PM to 10:15 PM?
   (b) Do you know how to accomplish (a)?
   If No, you need to study arithmetic.

4. (a) Do you sometimes want to work at a mortgage office?
   (b) Do you have the arithmetic preparation for this office?
   If No, you need to study arithmetic.

5. (a) Do you sometimes have to determine the time for traveling a certain distance at a certain speed?
   (b) Do you know how to accomplish (a)?
   If No, you need to study arithmetic.

6. (a) Do you sometimes compare a price per a certain quantity to a price per a different quantity?
   (b) Do you know how to accomplish (a)?
   If No, you need to study arithmetic.

7. (a) Do you sometimes have to determine the quantities of ingredients (e.g., salt, sugar, tomatoes) for food recipes for different parties?.
   (b) Do you know how to accomplish (a)?
   If No, you need to study arithmetic.

8. (a) Do you sometimes as a nurse have to calculate the correct dosage of a drug for a patient?
   (b) Do you have the arithmetic preparation to accomplish (a)?
   If No, you need to study arithmetic.

9. (a) Do you sometimes have to do calculations in chemistry?
   (b) Do you have the arithmetic background for (a)?
   If No, you need to study arithmetic.

10. (a) Do you sometimes want to work at the post office?
    (b) Do you have the arithmetic preparation for (a)?
    If No, you need to study arithmetic.

11. (a) Do you sometimes calculate employment and unemployment benefits?
    (b) Do you have the arithmetic preparation to accomplish (a)?
    If No, you need to study arithmetic.

12. (a) Do you sometimes work as a builder or a contractor?
    (b) Do you have the arithmetic preparation to accomplish (a)?
    If No, you need to study arithmetic.

13. (a) Do you sometimes compare interest rates and accompanying conditions?
    (b) Do you know how to accomplish (a)?
    If No, you need to study arithmetic.

14. (a) Do you sometimes have to divide profits among investors according to investment amounts?
    (b) Do you know how to accomplish (a)?
    If No, you need to study arithmetic

15. (a) Do you sometimes have to compare apartment or house prices?
    (b) Do you know how to accomplish (a)?
    If No you need to study arithmetic

16. (a) Do you sometimes have to deal with percents such as 20% of 60?
    (b) Do you know how to accomplish (a)?
    If No, you need to study arithmetic.

17. (a) Do you sometimes arrange items labeled with fractions or decimals in ascending or descending order?
    (b) Do you know how to accomplish (a)?
    If No, you need to study arithmetic

18. (a) Do you sometimes have to do estimations such as the area of the living room floor of your house?
    (b) Do you know how to accomplish (a)?
    If No, you need to study arithmetic.

19. (a) Do you sometimes have to
       work with fractions or decimals on the job?
    (b) Do you know how to accomplish (a)?
    If No, you need to study arithmetic.

20. (a) Do you sometimes have to prepare a budget?
    (b) Do you know how to accomplish (a)?
    If No, you need to study arithmetic.

21. (a) Do you sometimes want to work in finance?
    (b) Do you have the arithmetic preparation for (a)?
    If No, you need to study arithmetic.

22. (a) Do you sometimes want to go to Wall Street
       and do some trading?
    (b) Do you have the arithmetic preparation for (a)?
    If No, you need to study arithmetic.

23 (a) Do you sometimes work as a carpenter?
    (b) Do you have the arithmetic preparation for (a)?
    If No, you need to study arithmetic.

24. (a) Do you sometimes work as a bricklayer?
    (b) Do you have the arithmetic preparation for (a)?
    If No, you need to study arithmetic.

25. Do you sometimes take a civil service
    exam with a mathematics component?
    (b) Do you know how to accomplish (a)?
    If No, you need to study arithmetic.

26 (a) . Do you sometimes want to work
       as a bank teller?
    (b) Do you have the arithmetic
       background for (a)?
    If No, you need to study arithmetic.

27. (a) Do you sometimes have to work
       on personal finances?
    (b) Do you know how to accomplish (a)?
    If No, you need to study arithmetic.

28. (a) Do you sometimes have to
       compare college tuitions?
    (b) Do you know how to accomplish (a)?
    If No, you need to study arithmetic.

29. (a) Do you sometimes buy and sell
       goods for profit?
    (b) Do you have the arithmetic
       preparation for (a)?
    If No, you need to study arithmetic.

Read More

If you are very competent in **arithmetic** and you do not even have a formal college education, and you get a job as a banker, as a nurse, as an auto mechanic, as a construction worker, or a job in a finance company, or a mortgage company, you will be able to train quickly, and move up in rank in any company. If you are not competent in arithmetic, you will always guess, make mistakes, and make wrong decisions and you may be cheated, for example, if you cannot do the following problems correctly.

1. Suppose you and a friend invested $2,000 and $3,000 respectively, in a company, and you are to divide a profit of $1,500, how much is your share of this profit based on the amount you invested?

Answer: $600

2. Suppose you and a friend worked 8 hours and 12 hours respectively, as security guards at a trade show. If the total payment for the 20 hours is $260, what is your pay for the hours you worked?

Answer: $104

3. According to the decision criteria of a mortgage company, an applicant for a loan will qualify for the loan if the applicant's monthly income-expense ratio is greater than 1. If your monthly income is $5,000; and your monthly expenses are $4,900, do you qualify for the loan?   Answer: Yes

Some people look down upon the subject "arithmetic" as being so elementary and may hesitate to buy a book with such a title. However, most everyday life and business decisions are based on arithmetic. There are various levels of arithmetic from grade school, through high school, to college, and at the work place. For example, the following is an arithmetic problem.

If you deposited $30,000 in a savings account at an interest rate of 5%, and the interest is compounded annually for 2 years, what is the amount in your account after 2 years?

Answer: $33,075

We use arithmetic in making financial decisions, in accounting practices, in business management, in income statement preparation, and in determining the cost of goods sold, operating expenses,, net profit, and in managing accounts receivable, accounts payable, and cash flow.

1	2	3	4	5	6
$1 \times 1 = 1$	$2 \times 1 = 2$	$3 \times 1 = 3$	$4 \times 1 = 4$	$5 \times 1 = 5$	$6 \times 1 = 6$
$1 \times 2 = 2$	$2 \times 2 = 4$	$3 \times 2 = 6$	$4 \times 2 = 8$	$5 \times 2 = 10$	$6 \times 2 = 12$
$1 \times 3 = 3$	$2 \times 3 = 6$	$3 \times 3 = 9$	$4 \times 3 = 12$	$5 \times 3 = 15$	$6 \times 3 = 18$
$1 \times 4 = 4$	$2 \times 4 = 8$	$3 \times 4 = 12$	$4 \times 4 = 16$	$5 \times 4 = 20$	$6 \times 4 = 24$
$1 \times 5 = 5$	$2 \times 5 = 10$	$3 \times 5 = 15$	$4 \times 5 = 20$	$5 \times 5 = 25$	$6 \times 5 = 30$
$1 \times 6 = 6$	$2 \times 6 = 12$	$3 \times 6 = 18$	$4 \times 6 = 24$	$5 \times 6 = 30$	$6 \times 6 = 36$
$1 \times 7 = 7$	$2 \times 7 = 14$	$3 \times 7 = 21$	$4 \times 7 = 28$	$5 \times 7 = 35$	$6 \times 7 = 42$
$1 \times 8 = 8$	$2 \times 8 = 16$	$3 \times 8 = 24$	$4 \times 8 = 32$	$5 \times 8 = 40$	$6 \times 8 = 48$
$1 \times 9 = 9$	$2 \times 9 = 18$	$3 \times 9 = 27$	$4 \times 9 = 36$	$5 \times 9 = 45$	$6 \times 9 = 54$
$1 \times 10 = 10$	$2 \times 10 = 20$	$3 \times 10 = 30$	$4 \times 10 = 40$	$5 \times 10 = 50$	$6 \times 10 = 60$
$1 \times 11 = 11$	$2 \times 11 = 22$	$3 \times 11 = 33$	$4 \times 11 = 44$	$5 \times 11 = 55$	$6 \times 11 = 66$
$1 \times 12 = 12$	$2 \times 12 = 24$	$3 \times 12 = 36$	$4 \times 12 = 48$	$5 \times 12 = 60$	$6 \times 12 = 72$
$1 \times 13 = 13$	$2 \times 13 = 26$	$3 \times 13 = 39$	$4 \times 13 = 52$	$5 \times 13 = 65$	$6 \times 13 = 78$
$1 \times 14 = 14$	$2 \times 14 = 28$	$3 \times 14 = 42$	$4 \times 14 = 56$	$5 \times 14 = 70$	$6 \times 14 = 84$
$1 \times 15 = 15$	$2 \times 15 = 30$	$3 \times 15 = 45$	$4 \times 15 = 60$	$5 \times 15 = 75$	$6 \times 15 = 90$
$1 \times 16 = 16$	$2 \times 16 = 32$	$3 \times 16 = 48$	$4 \times 16 = 64$	$5 \times 16 = 80$	$6 \times 16 = 96$
$1 \times 17 = 17$	$2 \times 17 = 34$	$3 \times 17 = 51$	$4 \times 17 = 68$	$5 \times 17 = 85$	$6 \times 17 = 102$
$1 \times 18 = 18$	$2 \times 18 = 36$	$3 \times 18 = 54$	$4 \times 18 = 72$	$5 \times 18 = 90$	$6 \times 18 = 108$
$1 \times 19 = 19$	$2 \times 19 = 38$	$3 \times 19 = 57$	$4 \times 19 = 76$	$5 \times 19 = 95$	$6 \times 19 = 114$
$1 \times 20 = 20$	$2 \times 20 = 40$	$3 \times 20 = 60$	$4 \times 20 = 80$	$5 \times 20 = 100$	$6 \times 20 = 120$

7	8	9	10	11	12
$7 \times 1 = 7$	$8 \times 1 = 8$	$9 \times 1 = 9$	$10 \times 1 = 10$	$11 \times 1 = 11$	$12 \times 1 = 12$
$7 \times 2 = 14$	$8 \times 2 = 16$	$9 \times 2 = 18$	$10 \times 2 = 20$	$11 \times 2 = 22$	$12 \times 2 = 24$
$7 \times 3 = 21$	$8 \times 3 = 24$	$9 \times 3 = 27$	$10 \times 3 = 30$	$11 \times 3 = 33$	$12 \times 3 = 36$
$7 \times 4 = 28$	$8 \times 4 = 32$	$9 \times 4 = 36$	$10 \times 4 = 40$	$11 \times 4 = 44$	$12 \times 4 = 48$
$7 \times 5 = 35$	$8 \times 5 = 40$	$9 \times 5 = 45$	$10 \times 5 = 50$	$11 \times 5 = 55$	$12 \times 5 = 60$
$7 \times 6 = 42$	$8 \times 6 = 48$	$9 \times 6 = 54$	$10 \times 6 = 60$	$11 \times 6 = 66$	$12 \times 6 = 72$
$7 \times 7 = 49$	$8 \times 7 = 56$	$9 \times 7 = 63$	$10 \times 7 = 70$	$11 \times 7 = 77$	$12 \times 7 = 84$
$7 \times 8 = 56$	$8 \times 8 = 64$	$9 \times 8 = 72$	$10 \times 8 = 80$	$11 \times 8 = 88$	$12 \times 8 = 96$
$7 \times 9 = 63$	$8 \times 9 = 72$	$9 \times 9 = 81$	$10 \times 9 = 90$	$11 \times 9 = 99$	$12 \times 9 = 108$
$7 \times 10 = 70$	$8 \times 10 = 80$	$9 \times 10 = 90$	$10 \times 10 = 100$	$11 \times 10 = 110$	$12 \times 10 = 120$
$7 \times 11 = 77$	$8 \times 11 = 88$	$9 \times 11 = 99$	$10 \times 11 = 110$	$11 \times 11 = 121$	$12 \times 11 = 132$
$7 \times 12 = 84$	$8 \times 12 = 96$	$9 \times 12 = 108$	$10 \times 12 = 120$	$11 \times 12 = 132$	$12 \times 12 = 144$
$7 \times 13 = 91$	$8 \times 13 = 104$	$9 \times 13 = 117$	$10 \times 13 = 130$	$11 \times 13 = 143$	$12 \times 13 = 156$
$7 \times 14 = 98$	$8 \times 14 = 112$	$9 \times 14 = 126$	$10 \times 14 = 140$	$11 \times 14 = 154$	$12 \times 14 = 168$
$7 \times 15 = 105$	$8 \times 15 = 120$	$9 \times 15 = 135$	$10 \times 15 = 150$	$11 \times 15 = 165$	$12 \times 15 = 180$
$7 \times 16 = 112$	$8 \times 16 = 128$	$9 \times 16 = 144$	$10 \times 16 = 160$	$11 \times 16 = 176$	$12 \times 16 = 192$
$7 \times 17 = 119$	$8 \times 17 = 136$	$9 \times 17 = 153$	$10 \times 17 = 170$	$11 \times 17 = 187$	$12 \times 17 = 204$
$7 \times 18 = 126$	$8 \times 18 = 144$	$9 \times 18 = 162$	$10 \times 18 = 180$	$11 \times 18 = 198$	$12 \times 18 = 216$
$7 \times 19 = 133$	$8 \times 19 = 152$	$9 \times 19 = 171$	$10 \times 19 = 190$	$11 \times 19 = 209$	$12 \times 19 = 228$
$7 \times 20 = 140$	$8 \times 20 = 160$	$9 \times 20 = 180$	$10 \times 20 = 200$	$11 \times 20 = 220$	$12 \times 20 = 240$

13	14	15	16	17	18
$13 \times 1 = 13$	$14 \times 1 = 14$	$15 \times 1 = 15$	$16 \times 1 = 16$	$17 \times 1 = 17$	$18 \times 1 = 18$
$13 \times 2 = 26$	$14 \times 2 = 28$	$15 \times 2 = 30$	$16 \times 2 = 32$	$17 \times 2 = 34$	$18 \times 2 = 36$
$13 \times 3 = 39$	$14 \times 3 = 42$	$15 \times 3 = 45$	$16 \times 3 = 48$	$17 \times 3 = 51$	$18 \times 3 = 54$
$13 \times 4 = 52$	$14 \times 4 = 56$	$15 \times 4 = 60$	$16 \times 4 = 64$	$17 \times 4 = 68$	$18 \times 4 = 72$
$13 \times 5 = 65$	$14 \times 5 = 70$	$15 \times 5 = 75$	$16 \times 5 = 80$	$17 \times 5 = 85$	$18 \times 5 = 90$
$13 \times 6 = 78$	$14 \times 6 = 84$	$15 \times 6 = 90$	$16 \times 6 = 96$	$17 \times 6 = 102$	$18 \times 6 = 108$
$13 \times 7 = 91$	$14 \times 7 = 98$	$15 \times 7 = 105$	$16 \times 7 = 112$	$17 \times 7 = 119$	$18 \times 7 = 126$
$13 \times 8 = 104$	$14 \times 8 = 112$	$15 \times 8 = 120$	$16 \times 8 = 128$	$17 \times 8 = 136$	$18 \times 8 = 144$
$13 \times 9 = 117$	$14 \times 9 = 126$	$15 \times 9 = 135$	$16 \times 9 = 144$	$17 \times 9 = 153$	$18 \times 9 = 162$
$13 \times 10 = 130$	$14 \times 10 = 140$	$15 \times 10 = 150$	$16 \times 10 = 160$	$17 \times 10 = 170$	$18 \times 10 = 180$
$13 \times 11 = 143$	$14 \times 11 = 154$	$15 \times 11 = 165$	$16 \times 11 = 176$	$17 \times 11 = 187$	$18 \times 11 = 198$
$13 \times 12 = 156$	$14 \times 12 = 168$	$15 \times 12 = 180$	$16 \times 12 = 192$	$17 \times 12 = 204$	$18 \times 12 = 216$
$13 \times 13 = 169$	$14 \times 13 = 182$	$15 \times 13 = 195$	$16 \times 13 = 208$	$17 \times 13 = 221$	$18 \times 13 = 234$
$13 \times 14 = 182$	$14 \times 14 = 196$	$15 \times 14 = 210$	$16 \times 14 = 224$	$17 \times 14 = 238$	$18 \times 14 = 252$
$13 \times 15 = 195$	$14 \times 15 = 210$	$15 \times 15 = 225$	$16 \times 15 = 240$	$17 \times 15 = 255$	$18 \times 15 = 270$
$13 \times 16 = 208$	$14 \times 16 = 224$	$15 \times 16 = 240$	$16 \times 16 = 256$	$17 \times 16 = 272$	$18 \times 16 = 288$
$13 \times 17 = 221$	$14 \times 17 = 238$	$15 \times 17 = 255$	$16 \times 17 = 272$	$17 \times 17 = 289$	$18 \times 17 = 306$
$13 \times 18 = 234$	$14 \times 18 = 252$	$15 \times 18 = 270$	$16 \times 18 = 288$	$17 \times 18 = 306$	$18 \times 18 = 324$
$13 \times 19 = 247$	$14 \times 19 = 266$	$15 \times 19 = 285$	$16 \times 19 = 304$	$17 \times 19 = 323$	$18 \times 19 = 342$
$13 \times 20 = 260$	$14 \times 20 = 280$	$15 \times 20 = 300$	$16 \times 20 = 320$	$17 \times 20 = 340$	$18 \times 20 = 360$

19	20	21	22	23	24
$19 \times 1 = 19$	$20 \times 1 = 20$	$21 \times 1 = 21$	$22 \times 1 = 22$	$23 \times 1 = 23$	$24 \times 1 = 24$
$19 \times 2 = 38$	$20 \times 2 = 40$	$21 \times 2 = 42$	$22 \times 2 = 44$	$23 \times 2 = 46$	$24 \times 2 = 48$
$19 \times 3 = 57$	$20 \times 3 = 60$	$21 \times 3 = 63$	$22 \times 3 = 66$	$23 \times 3 = 69$	$24 \times 3 = 72$
$19 \times 4 = 76$	$20 \times 4 = 80$	$21 \times 4 = 84$	$22 \times 4 = 88$	$23 \times 4 = 92$	$24 \times 4 = 96$
$19 \times 5 = 95$	$20 \times 5 = 100$	$21 \times 5 = 105$	$22 \times 5 = 110$	$23 \times 5 = 115$	$24 \times 5 = 120$
$19 \times 6 = 114$	$20 \times 6 = 120$	$21 \times 6 = 126$	$22 \times 6 = 132$	$23 \times 6 = 138$	$24 \times 6 = 144$
$19 \times 7 = 133$	$20 \times 7 = 140$	$21 \times 7 = 147$	$22 \times 7 = 154$	$23 \times 7 = 161$	$24 \times 7 = 168$
$19 \times 8 = 152$	$20 \times 8 = 160$	$21 \times 8 = 168$	$22 \times 8 = 176$	$23 \times 8 = 184$	$24 \times 8 = 192$
$19 \times 9 = 171$	$20 \times 9 = 180$	$21 \times 9 = 189$	$22 \times 9 = 198$	$23 \times 9 = 207$	$24 \times 9 = 216$
$19 \times 10 = 190$	$20 \times 10 = 200$	$21 \times 10 = 210$	$22 \times 10 = 220$	$23 \times 10 = 230$	$24 \times 10 = 240$
$19 \times 11 = 209$	$20 \times 11 = 220$	$21 \times 11 = 231$	$22 \times 11 = 242$	$23 \times 11 = 253$	$24 \times 11 = 264$
$19 \times 12 = 228$	$20 \times 12 = 240$	$21 \times 12 = 252$	$22 \times 12 = 264$	$23 \times 12 = 276$	$24 \times 12 = 288$
$19 \times 13 = 247$	$20 \times 13 = 260$	$21 \times 13 = 273$	$22 \times 13 = 286$	$23 \times 13 = 299$	$24 \times 13 = 312$
$19 \times 14 = 266$	$20 \times 14 = 280$	$21 \times 14 = 294$	$22 \times 14 = 308$	$23 \times 14 = 322$	$24 \times 14 = 336$
$19 \times 15 = 285$	$20 \times 15 = 300$	$21 \times 15 = 315$	$22 \times 15 = 330$	$23 \times 15 = 345$	$24 \times 15 = 360$
$19 \times 16 = 304$	$20 \times 16 = 320$	$21 \times 16 = 336$	$22 \times 16 = 352$	$23 \times 16 = 368$	$24 \times 16 = 384$
$19 \times 17 = 323$	$20 \times 17 = 340$	$21 \times 17 = 357$	$22 \times 17 = 374$	$23 \times 17 = 391$	$24 \times 17 = 408$
$19 \times 18 = 342$	$20 \times 18 = 360$	$21 \times 18 = 378$	$22 \times 18 = 396$	$23 \times 18 = 414$	$24 \times 18 = 432$
$19 \times 19 = 361$	$20 \times 19 = 380$	$21 \times 19 = 399$	$22 \times 19 = 418$	$23 \times 19 = 437$	$24 \times 19 = 456$
$19 \times 20 = 380$	$20 \times 20 = 400$	$21 \times 20 = 420$	$22 \times 20 = 440$	$23 \times 20 = 460$	$24 \times 20 = 480$

If you can remember a needed information, you make decisions faster, you work faster, and you are more productive.

Yes, you can memorize them.
Squares of Natural Numbers

$1 \times 1 = 1$	$26 \times 26 = 676$
$2 \times 2 = 4$	$27 \times 27 = 729$
$3 \times 3 = 9$	$28 \times 28 = 784$
$4 \times 4 = 16$	$29 \times 29 = 841$
$5 \times 5 = 25$	$30 \times 30 = 900$
$6 \times 6 = 36$	$31 \times 31 = 961$
$7 \times 7 = 49$	$32 \times 32 = 1024$
$8 \times 8 = 64$	$33 \times 33 = 1089$
$9 \times 9 = 81$	$34 \times 34 = 1156$
$10 \times 10 = 100$	$35 \times 35 = 1225$
$11 \times 11 = 121$	$36 \times 36 = 1296$
$12 \times 12 = 144$	$37 \times 37 = 1369$
$13 \times 13 = 169$	$38 \times 38 = 1444$
$14 \times 14 = 196$	$39 \times 39 = 1521$
$15 \times 15 = 225$	$40 \times 40 = 1600$
$16 \times 16 = 256$	$41 \times 41 = 1681$
$17 \times 17 = 289$	$42 \times 42 = 1764$
$18 \times 18 = 324$	$43 \times 43 = 1849$
$19 \times 19 = 361$	$44 \times 44 = 1936$
$20 \times 20 = 400$	$45 \times 45 = 2025$
$21 \times 21 = 441$	$46 \times 46 = 2116$
$22 \times 22 = 484$	$47 \times 47 = 2209$
$23 \times 23 = 529$	$48 \times 48 = 2304$
$24 \times 24 = 576$	$49 \times 49 = 2401$
$25 \times 25 = 625$	$50 \times 50 = 2500$

Symbols for writing numbers: the **digits** or **numerals**
The **ten basic digits** for writing numbers (decimal system) are 0, 1, 2, 3, 4, 5, 6, 7, 8, and 9.

Natural Numbers = {1, 2, 3, 4, 5, 6, 7, 8, 9, 10, 11, 12, 13,...}
The natural numbers are also known as the counting numbers or the positive integers.

Whole numbers = {0, 1, 2, 3, 4, 5, 6, 7, 8, 9, 10, 11, 12, 13,...}

Integers = {...,-7, -6, -5. -4, -3, -2, -1, 0, 1, 2, 3, 4, 5, 6, 7,...}

Rational number : A number which can be written as the ratio of two integers.

Irrational number : A number which **cannot** be written as the ratio of two integers.

Example of a number showing the **place-value** relationships:

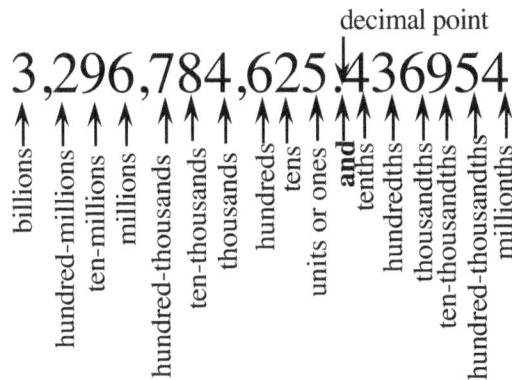

decimal point

3,296,784,625.436954

billions
hundred-millions
ten-millions
millions
hundred-thousands
ten-thousands
thousands
hundreds
tens
units or ones
and
tenths
hundredths
thousandths
ten-thousandths
hundred-thousandths
millionths

The above number is read " three billion, two hundred ninety-six million, seven hundred eighty-four thousand, six hundred twenty-five **and** four hundred thirty-six thousand, nine hundred fifty-four millionths".

Conversion Factors for Measurements

American System (British System)   Interconversion (Factors)   Metric System

↓ (American System)   ↓ (Interconversion)   ↓ (Metric System)

Some " **bridges**" for converting from one system to the other

1 kilometer (km)	$= 10^3$ m = 1000 m
1 hectometer (hm)	$= 10^2$ m = 100 m
1 dekameter (dam)	$= 10^1$ m = 10 m
1 meter (m)	$= 10^0$ m = 1m
1 decimeter (dm)	$= 10^{-1}$ m = 0.1m
1 centimeter (cm)	$= 10^{-2}$ m = 0.01m
1 millimeter (mm)	$= 10^{-3}$ m = 0.001m

Length

12 inches (in) = 1 foot (ft.)
3 feet (ft.)   = 1 yard (yd)
5280 feet   = 1 mile (mi)
1760 yards   = 1 mile

1 in. = 2.54 cm
1 yd = 0.9144 m
1 mi = 1.61 km
1 km = 0.62 mi

Some " **bridges**" for converting from one system to the other

1 kilogram (kg)	$= 10^3$ g = 1000 g
1 hectogram (hg)	$= 10^2$ g = 100 g
1 dekagram (dag)	$= 10^1$ g = 10 g
1 gram (g)	$= 10^0$ g = 1 g
1 decigram (dg)	$= 10^{-1}$ g = 0.1g
1 centigram (cg))	$= 10^{-2}$ g = 0.01g
1 milligram (mg)	$= 10^{-3}$ g = 0.001g

Mass

1 *l*b   = 16 oz
1 ton = 2000 *l*b
1 long ton = 2240 *l*b

1 kg = 2.2 *l*b
1 *l*b = 454 g
1 oz = 28.4 g = 16 drams
1 ton = 0.9072 metric ton

Some " **bridges**" for converting from one system to the other

1 kiloliter (k*l*)	$= 10^3$ *l* = 1000 *l*
1 hectoliter (h*l*)	$= 10^2$ *l* = 100 *l*
1 dekaliter (da*l*)	$= 10^1$ *l* = 10 *l*
1 liter (*l*)	$= 10^0$ *l* = 1 *l*
1 deciliter (d*l*)	$= 10^{-1}$ *l* = 0.1*l*
1 centiliter (c*l*)	$= 10^{-2}$ *l* = 0.01*l*
1 milliliter (m*l*)	$= 10^{-3}$ *l* = 0.001*l*

Volume

16 fluid oz (fl-oz)= 1 pint (pt)
2 pints (pt) = 1 quart (qt)
4 quarts   = 1 gallon (gal)

1 liter (*l*) = 1.057 qt
1 gal = 3.785 *l*
1 liter = 2.1 pt
1 pt = .473 *l*

Mnemonic device (metric system)

Must remember the following (metric system:)

100 cm   = 1 m
1000 m = 1 km

1000 mg = 1 g
1000 g   = 1 kg

1000 m*l* = 1 *l*
1 m*l*   = 1 cc = 1cm^3
1000 cc = 1 *l*

k – ilo – 10^3
h – ecto – 10^2
d – eka – 10^1
d – eci – 10^{-1}
c – enti – 10^{-2}
m – illi – 10^{-3}

Say the following aloud:
Step 1: First go down vertically as kei-eitch-dii-dii-see-em, then Step 2
Step 2: Kilo-hecto-deka-deci-centi-milli, and then note how the powers decrease vertically downwards.
Examples: 1 Kilometer = 10^3meter; 1 milligram =10^{-3}gram;
1 centimeter =10^{-2} meter =$\frac{1}{100}$ meter --->100 centimeters = 1 meter.

FREMPONG'S STEP-BY-STEP SERIES IN MATHEMATICS

Final Exam Review: Arithmetic

Second Edition

Excellent for Exam Preparation

A.A.FREMPONG

Final Exam Review: Arithmetic (Mother of Mathematics)

Copyright © 2018, 2016, 2011, 2010, 2006, 1999 by A. A. Frempong

Published by Yellowtextbooks.com/Finalexamsreview.com. N.Y.

ISBN 978-1-946485-22-9

Printed in the United States of America

In Memory of My Parents

Mom:
She was a devoted mother, sharing, kind, kinder to strangers and generous to a fault. She never cursed, she never hated; she never cheated, and she never envied. She never lied, and she never got angry. Once, she nursed an almost dying stranger renting a room in her house back to good health to the extent that the relatives of this renter later travelled one hundred miles just to thank mom. She was always peaceloving and forever forgiving.
An angel once lived on this earth to serve others.

Dad:
A great dad, kind, generous and forgiving. He emphasized and was an example of both formal education and self-education. A veterinarian, a bacteriologist, an Associate of the Institute of Medical Laboratory Technology (UK), a Fellow of the Royal Society of Health (UK); an incorruptible civil servant; his book on ticks has always inspired me to write whenever the need arises.

v

NOTE TO THE STUDENT

All learning is self-teaching. Everybody else can only help the student to learn.

You should always aim at getting an "A" on every exam, and especially on the final exam, since some instructors weigh the final exam more than a class exam. Do not say " I just want to pass." Show all work even if the exam is of the multiple-choice-answer type. If you did not do well on the class exams, and show all work, and excel in choosing the correct answers, your instructor would be easily convinced that you did not cheat on the final exam.

Since the final exam may be given a few days (or the next day) after the last class meeting, do not wait to begin preparing for the final exam after the last class meeting; but rather, you should begin preparing for the final exam, at least, three weeks (or better, four weeks) before the final exam date.
Confer with your instructor to obtain all the topics that would be covered by the end of the semester or term. In addition to the textbook and the class notes, refer to other mathematics textbooks, and notebooks of friends and family members.

Begin to master the definitions and the solutions of the sample problems.
To master a sample problem, study the sample problem and its solution process first, and then try to do the same problem without looking at the solution process. If you are able to repeat the solution process successfully, you have mastered the sample problem. If you can solve similar problems without any reference to this book or any other source, you have also mastered similar problems. For some problems, two or more methods are presented. Read the various methods and decide which methods you would like to remember; but always be aware of the existence of the other methods, in case the need arises. After having mastered the sample problems, try the exercise problems at the end of the lesson. You may refer back and forth to the solved problems when you do not remember how to proceed.
Finally, **repeat** all of the above from time to time. You will experience that you work faster on the old topics than on the new topics. It must be emphasized that **repetition** is very important.

As a reminder, in any book, do not dwell on the few inadvertent errors you may find, but rather concentrate on what is useful to you.

If you are working part-time or full-time, it would be helpful if you are on vacation from work during the last two weeks of the semester. If you are offered a new employment, let the first day on your new job be after the last final exam.

For this book to be useful, as review for exams, it is **important** to **understand, remember, apply**, and remember the material covered.

Wishing you Good Luck on the final exam.
A. A. Frempong

Books in the series by the author**: Integrated Arithmetic**; Elementary Algebra; Intermediate Algebra, Elementary Mathematics; Intermediate Mathematics; Elementary & Intermediate Mathematics (combined); College Algebra; College Trigonometry; College Algebra & Trigonometry and Calculus 1 & 2.

PREFACE

A new approach to learning and reviewing the basic topics in arithmetic is covered. This book is aimed at those students who have studied arithmetic in the past or recently. It is therefore appropriate, within the first few lessons, to re-exposethese students to fractions, decimals, and the percent symbol.

In the first chapter, the basic definitions of numbers, fractions, decimals, ratio, proportion, and percents, as well as the relationships between these terms are covered. In subsequent chapters, each topic is then treated in detail, regarding the various arithmetic operations.

The overall style of this book is informal, because the main objective is to communicate. A step-by-step approach is used throughout the book.

This book, is a spin-off of the "mother" edition , Elementary Mathematics by the author. The mother edition covers both arithmetic and elementary algebra and it is in its fifth edition.
One topic that has been covered very comprehensively is "Ratio and Proportion" (p.53-88), since this topic is the bread-and-butter arithmetic of every day life, and every educated person should be at ease with this topic. The ratios include ratios involving three or more terms; and the proportion problems cover both direct
and inverse proportion, as well as compound proportion. Various methods (up to five methods) for solving various proportion problems are covered. Both arithmetic and algebraic methods have been covered. If one has no algebraic background, one can skip the algebraic methods. However, there is material for some basic algebra in the Appendix (p.135-163).

This book can also be used for short term programs such as mini-sessions, immersion programs, and workshops, as well as for review.

I gratefully acknowledge the help and encouragement of students, colleagues, and friends.

A.A. Frempong
New York

CONTENTS

CHAPTER 6
PROPORTION (VARIATION)

CHAPTER 7   **89**

CHAPTER 8   **104**
Miscellaneous Topics

CLASS TESTS (Arithmetic)   **122**

APPENDIX A
Elementary Algebra
Signed Numbers and Real Number Operations

APPENDIX B   **148**

APPENDIX C
First Degree Equations Containing One Variable   **155**
(Simple Linear Equations)

APPENDIX D
About Measurements   **165**
Standard Unit, Error, Rounding-off Numbers, Significant Digits, Scientific Notation

INDEX   **177**

CHAPTER 1

Lesson 1: **Place-Value Relationships; Basic Definitions; Terminology; and Types of Numbers**
Lesson 2: **Introduction to Fractions, Decimals, Ratio and Proportion Percent Symbol and Some Interpretations**

Lesson 1

Symbols for Writing Numbers

The symbols for writing numbers are called the **digits** or **numerals**. The **ten basic digits** for writing numbers, in the decimal system, are 0, 1, 2, 3, 4, 5, 6, 7, 8, and 9. Using these digits we can write all the numbers that we cover in arithmetic.

Place-value relationships

When a number is written using digits, each digit has a value according to its position in the number. For example, in 45, the 4 represents four tens (or 4×10) and we say that the 4 is in the tens place. However, in the number 415, the 4 represents four hundreds (or 4×100) and we say that the 4 is in the hundreds place. Below, we present a number showing the place-value-relationships.

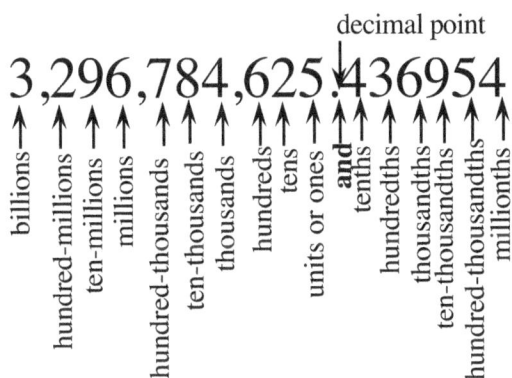

decimal point

$$3,296,784,625.436954$$

billions → hundred-millions → ten-millions → millions → hundred-thousands → ten-thousands → thousands → hundreds → tens → units or ones → **and** → tenths → hundredths → thousandths → ten-thousandths → hundred-thousandths → millionths

The above number is read " three billion, two hundred ninety-six million, seven hundred eighty-four thousand, six hundred twenty-five **and** four hundred thirty-six thousand, nine hundred fifty-four millionths".

We shall briefly look at the different kinds of numbers that we deal with in mathematics, aided by a number flow chart below, and then we will branch-off to the numbers that we deal with in arithmetic.

From the Entire Picture of Numbers to Arithmetic

The basic elements we deal with in our study of mathematics are numbers. It is important that we obtain a good understanding of the types of numbers we deal with in mathematics. It is also important that we are able to distinguish between the different kinds of numbers and their associated terminology. A very good grasp of the terminology will help us read and understand subsequent material much more quickly than otherwise.

As shown in the number flow chart below, all the numbers we deal with in mathematics can be divided into two main sets, namely the set of **real numbers** and the set of **non-real numbers**. (We must note that the terms "real", "non-real", and similar terms are only names we use in mathematics to distinguish between numbers and we must note that the real numbers are **not** more real than the non-real numbers (literally). These terms are only names for convenient distinction between some numbers.) The real numbers are also divided into two main sets, namely the **rational numbers** and the **irrational numbers.**

The rational numbers are further subdivided into **integers** (strictly integers) and **fractions** (strictly fractions).   The integers consist of the **negative integers, zero,** and the **positive integers** (natural numbers); and the positive integers consists of the **prime numbers** and the **composite numbers**.

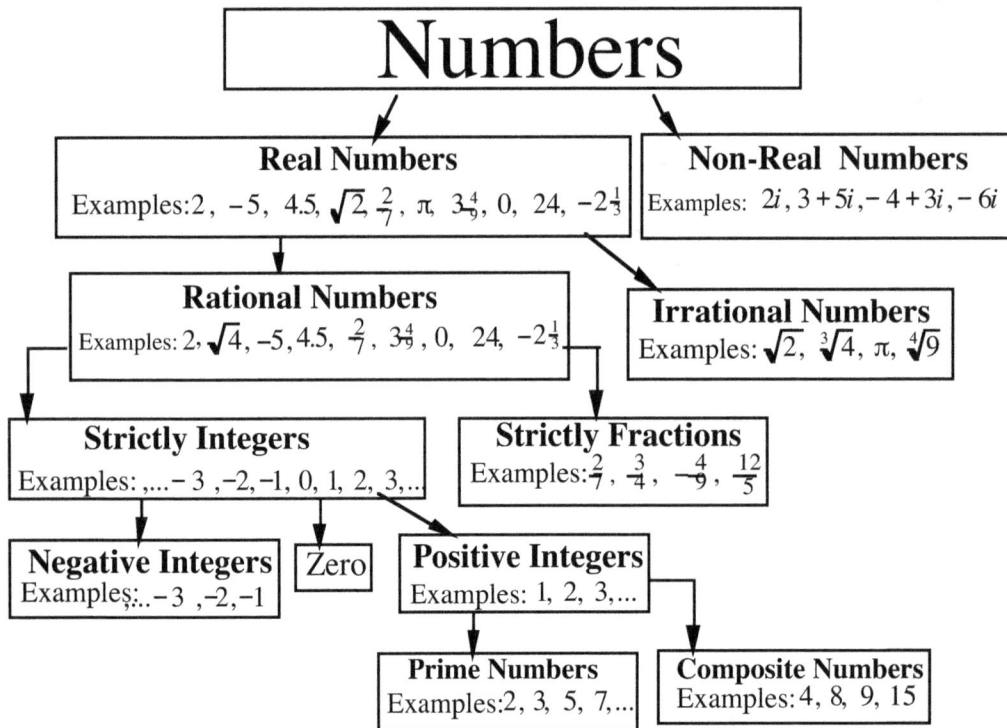

Numbers

Real Numbers

Examples: $2, -5, 4.5, \sqrt{2}, \frac{2}{7}, \pi, 3\frac{4}{9}, 0, 24, -2\frac{1}{3}$

Non-Real Numbers

Examples: $2i, 3+5i, -4+3i, -6i$

Rational Numbers

Examples: $2, \sqrt{4}, -5, 4.5, \frac{2}{7}, 3\frac{4}{9}, 0, 24, -2\frac{1}{3}$

Irrational Numbers

Examples: $\sqrt{2}, \sqrt[3]{4}, \pi, \sqrt[4]{9}$

Strictly Integers

Examples: $,\ldots -3, -2, -1, 0, 1, 2, 3,\ldots$

Strictly Fractions

Examples: $\frac{2}{7}, \frac{3}{4}, \frac{4}{9}, \frac{12}{5}$

Negative Integers

Examples: $\ldots -3, -2, -1$

Zero

Positive Integers

Examples: $1, 2, 3,\ldots$

Prime Numbers

Examples: $2, 3, 5, 7,\ldots$

Composite Numbers

Examples: $4, 8, 9, 15$

Number Flow Chart

In **arithmetic**, we will deal mainly with **non-negative real numbers** (i.e., zero, and positive real numbers which include the positive integers, positive fractions, mixed numbers and decimals). We shall now define some terms that we use very often to communicate in arithmetic.

Lesson 1: Place-Value Relationships; Basic Definitions; Terminology; Types of Numbers

We define a **set of numbers** as a well-defined collection of numbers.

The set of the **natural numbers** consists of the numbers $1, 2, 3, 4, 5, 6, 7, 8, 9, 10, 11, 12, 13,...,$ If we know a natural number, to obtain the next natural number we add 1. The smallest natural number is 1, but we do not know the largest natural number, since given even a very large natural number, we can always obtain the next natural number by adding 1. The natural numbers are also known as the **counting numbers** or the **positive integers**.

The set of **whole numbers** consists of the numbers $0, 1, 2, 3, 4, 5, 6, 7, 8, 9, 10, 11, 12, 13,...,$ If we know a whole number, to obtain the next whole number we add 1. The smallest whole number is 0, but we do not know the largest whole number, since given even a very large whole number, we can always obtain the next whole number by adding 1.

Even though we do not cover a set of numbers called **negative integers** in arithmetic, we briefly mention them here for completeness. If we take the opposites of the set of natural numbers also called positive integers, we obtain a set of numbers called **negative integers** (such as -1, -2, and -12).. If we combine the whole numbers with the negative integers we obtain a set of numbers called the **integers**. The set of integers therefore consists of the set of numbers $...,-6, -5. -4,$ $-3, -2, -1, 0, 1, 2, 3, 4, 5, 6,...$ (The three dots preceding the -6 on the left indicates that the numbers continue to decrease to the left and the three dots after the 6 on the right indicates that the numbers continue to increase to the right)

Ratio: The ratio a is b is the fraction $\frac{a}{b}$. **Example:** The ratio 3 is to 4 is the fraction $\frac{3}{4}$.

Rational number: A rational number (a fraction) is a number which **can** be written as the ratio of two integers. The word **rational** pertains to the word **ratio.**

Examples are (a) $\frac{2}{3}$;   (b) $\frac{1}{5}$;   (c)   4   (since $4 = \frac{4}{1}$)

(d)   0   (since $0 = \frac{0}{7} = \frac{0}{3}$ ... or $0 = \frac{0}{b}$, where b is an integer and b $\neq$ 0)

(e) $\sqrt{4}$ (because $\sqrt{4} = 2 = \frac{2}{1}$)

A rational number can also be written either as a terminating decimal or as a repeating decimal.

Examples of terminating decimals: $\frac{1}{4} = .25;$ $\frac{13}{2} = 6.5;$ $\frac{37}{8} = 4.625$.

Examples of repeating decimals: $\frac{1}{3} = .333...$ or $.\overline{3}$ and $\frac{1}{6} = .1666...$ or $.1\overline{6}$; $\frac{2}{3} = .66...$ or $.\overline{6}$.

Note the bar (vinculum) placed over the repeating digit or block of digits.
We may also regard a terminating decimal as non-terminating if we attach zeros to the right of the decimal. Examples are .25 = .250000...,   .5 = .5000...

Other types of numbers we cover in arithmetic are fractions, mixed numbers, and decimals.
There is another set of numbers, namely the irrational numbers. We define an **irrational number** as a number which **cannot** be written as the ratio of two integers. However, we can approximate irrational numbers as closely as we wish by rational numbers or decimals.

Examples of irrational numbers are $\sqrt{2}$, $\sqrt[3]{4}$, and π (pi). We can for example, approximate $\sqrt{2}$ by 1.414, and π by $\frac{22}{7}$ or 3.142.

When written in decimal form, an irrational number is non-repeating and non-terminating.

If we combine the natural numbers, the fractions, decimals, irrational numbers, the negatives of these numbers, and zero , we obtain the set of **real numbers**. Simply, the real numbers consists of the rational numbers and the irrational numbers.

Lesson 1: Place-Value Relationships; Basic Definitions; Terminology; Types of Numbers

We can represent numbers by points on a horizontal line called the **real number line** (Fig.1)

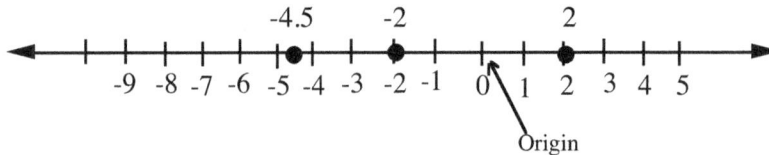

Figure 1

The **real number line** is a horizontal straight line with equally spaced intervals as in Figure 1 above. We label a point called the origin, 0 (zero). Points to the right of the origin are labeled positive and points to the left of the origin are labeled negative. The numbers increase as one moves from the left to the right on the real number line. Roughly speaking, a real number is a number that can be represented by a point on the real number line. The real numbers consists of the integers, fractions, mixed numbers, decimals, and radicals. In Figure 1, if the real numbers, -4.5, -2, and 2 are of interest, we can represent them by the dots shown. Every point on this line is associated with a real number; and every real number is associated with a point on this line. We can also say that the set of real numbers consists of the signed numbers and zero.

The arithmetic number line
Since in arithmetic, we deal with the numbers from zero to the right on the real number line, we can consider the **arithmetic number line** as the part of the real number line beginning from zero and extending to the right.

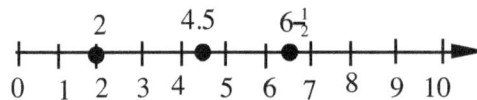

Summary for Some Number Terminology

Positive integers, or natural numbers or counting numbers $= \{1, 2, 3, 4,...\}$

Negative integers $= \{...,-4, -3, -2, -1\}$

Non-negative integers or **whole numbers** $= \{0, 1, 2, 3, 4,...\}$

Non-positive integers $= \{...,-4, -3, -2, -1, 0\}$

Non-negative real numbers: Examples are $0, 4, 7.5, 3\frac{1}{2}, \frac{1}{4}, 6,$ and $\sqrt{2}$

Non-positive real numbers: Examples are $-\sqrt{11}, -4, -3\frac{1}{2}, -2, -1, -\frac{1}{4}, .126,$ and 0.

We must understand thoroughly the above terms because they will be used over and over in the future. Anytime we meet any of these terms, we should try to form a quick mental picture of representative examples.

Lesson 1 Exercises

1. Define the following:
   Natural numbers, counting numbers, positive integers, whole numbers , negative integers, integers, rational number

2. From memory, draw the number flow chart in Lesson 1.

Lesson 2

Introduction to Fractions, Decimals, Ratio and Proportion; Percent Symbol and Some Interpretations

Fractions (In **chapter 3**, we cover the various operations on fractions and mixed numbers)

Fraction: A fraction is a number representing one or more of the equal parts into which a whole (thing) has been divided.

Unit fraction   One part of a whole which has been divided (or broken) into equal parts.

The numerator of a unit fraction is 1. Examples of unit fractions are $\frac{1}{2}, \frac{1}{8}$ and $\frac{1}{100}$.

Terms and Meaning of a Fraction

Example   In the fraction, $\frac{2}{3}$, the 2 and 3 are called the terms of the fraction; the 2 is called the numerator and the 3 is called the denominator.

In a fraction, the numerator indicates how many of the unit fraction the fraction represents; the denominator indicates into how many equal parts a whole has been divided; the denominator also indicates the size of the unit fraction: the smaller the denominator, the larger the size, and the larger the denominator, the smaller the size of the unit fraction.

Note that in addition to representing parts of a whole, a **fraction** is also used to express **division**, and also to **compare** numbers.

Types of Fractions

Proper Fraction: A fraction in which the numerator is less than the denominator.

Examples are $\frac{4}{5}, \frac{2}{7}$ and $\frac{34}{45}$. (A proper fraction is always less than 1)

Improper Fraction: A fraction in which the numerator is greater than or equal to the denominator.

Examples are $\frac{8}{5}, \frac{9}{3}$ and $\frac{7}{7}$. (An improper fraction is always greater than or equal to 1)

Mixed Number: A mixed number consists of a whole number part and a proper fraction part. We obtain a mixed number from an improper fraction when the denominator does not divide the numerator exactly (i.e. without a remainder).

Converting an Improper Fraction to a Mixed Number

Example Convert $\frac{91}{17}$ to a mixed number.

Solution $\frac{91}{17}$ = 5 remainder 6 (using long division)

$$\begin{array}{r} 5 \\ 17\overline{)91} \\ -85 \\ \hline 6 \end{array}$$

Therefore, $\frac{91}{17} = 5\frac{6}{17}$

Converting a Mixed Number to an Improper fraction

Example Convert $6\frac{3}{4}$ to an improper fraction.

Solution   Step 1: $6 \times 4 = 24$

Step 2: $24 + 3 = 27$

Step 3: $6\frac{3}{4} = \frac{27}{4}$ (Normally, we do the conversion mentally and show only Step 3.)

Decimal Fractions and Decimals

(In **Chapter 4**, we cover the various operations on decimals)

Powers of 10

Examples:
$$10^1 = 10$$
$$10^2 = 100$$
$$10^3 = 1,000$$
$$10^4 = 10,000$$

A **decimal fraction** is a fraction whose denominator is a power of 10.

Examples are $\frac{3}{10}$, $\frac{5}{100}$, and $\frac{23}{1000}$ . .

A **decimal** is the symbol for an equivalent decimal fraction.

Examples: $\frac{3}{10} = .3$ ($\frac{3}{10}$ is the decimal fraction; .3 is the decimal)

$\frac{5}{100} = .05$

$\frac{23}{10000} = .0023$

Ratio and Proportion

(In **chapters 5 & 6**, we cover ratio and proportion problems)

The **ratio** of two quantities a and b (written $a{:}b$) is the fraction $\frac{a}{b}$ (i.e., is the first quantity (mentioned) divided by the second quantity (mentioned)). The quantities a and b are called the terms of the ratio.

Proportion : A proportion is a statement that two ratios are equal.

The proportion "a is to b as c is to d " (or $a{:}b{::}c{:}d$) is the equality

$$\frac{a}{b} = \frac{c}{d} \tag{1}$$

In the above proportion, a and d are called the extremes and b and c are called the means of the proportion.

We may also define a proportion as a mathematical statement indicating the equality between two **equivalent fractions.**   Example $\frac{2}{3} = \frac{4}{6}$ is a proportion

Percent symbol " % " and some interpretations

(In **chapter 7**, we cover problems involving percent (%))

1. **Over hundred**: For example, 20% means $\frac{20}{100}$.   (Twenty over hundred or 20 divided by 100)

2. **For each hundred**: For example, a savings account with a 5% interest rate pays the depositor $5 for each $100 in the account. (Five for each hundred)

3. **Hundredths:** For example, 20% means $\frac{20}{100}$ or .20   (Twenty hundredths)

4. **As a number out of 100**: For example, a grade of 80% on a test means a student got 80 points out of 100 points.

Converting a decimal (or any fraction) **to percent** 7

Procedure: Multiply by 100%. (i.e., multiply by 100 and attach the percent symbol "%".

To change a decimal to percent, move the decimal point two places to the right and attach the percent symbol. (Note that moving the decimal point two places to the right is equivalent to multiplying by 100)

Example 1 Convert .74 to percent.

Solution   .74 = 74% .

Example 2 Convert .008 to percent.

Solution   .008 = .8%

Example 3 Convert 1 to percent.

 Solution   1 = 100%

Example 4 Convert 12 to percent.

Solution   12. = 1200%

Example 5   Change $\frac{1}{4}$ to percent

Method 1 $\frac{1}{4} \times \frac{100\%}{1} = 25\%$

Method 2 $\frac{1}{4} = .25 = .25 \times 100\%$ (Changing to decimal first, and then changing to percent)

Converting a percent to a decimal

Procedure: Divide by 100%. (i.e., drop the percent symbol "%" and divide by 100) or apply the meaning of the percent symbol to change the percent to a decimal fraction and then easily to a decimal.

To change a percent to a decimal, drop the percent symbol and move the decimal point two places to the left .   (**Note** that moving the decimal point two places to the left is equivalent to dividing by 100)

Example 1   74% = .74   (or 74% = $\frac{74}{100}$ = .74)

Example 2   .8% = .008   (or .8% = $\frac{.8}{100}$ = .008)

Example 3   145% = 1.45

Example 4   14.5% = .145

Example 5   .145% = .00145

Note above: Attaching the % symbol is equivalent to dividing by 100, and dropping the % symbol is equivalent to multiplying by 100.

Lesson 2 Exercises

A. 1 Convert to mixed numbers: (a). $\frac{22}{7}$;   (b). $\frac{17}{5}$;   (c) $\frac{79}{18}$.

2 Verify the answers in **A** by converting your answers back to improper fractions.

3. Change to improper fractions. (a) $8\frac{5}{6}$;   (b). $9\frac{3}{8}$

Answers: **1. (a).** $3\frac{1}{7}$; **(b)**. $3\frac{2}{5}$; (c) $4\frac{7}{18}$; **3**: **(a)** $\frac{53}{6}$; **(b)**. $\frac{75}{8}$.

B. Write as decimals: (a) $\frac{36}{1000}$; (b) $\frac{40}{100}$; (c) $\frac{9}{100}$

Answers (a) .036 ;   (b).40 ;   (c) .09

C Change each of the following to a fraction in its lowest terms:

   (a) .64 ;   (b) 6.4 ;   (c) 0.036 ;   (d) 1.36 ;   (e) .63;   (f) 0.0825

Answers (a) $\frac{16}{25}$; (b) $\frac{32}{5}$; (c) $\frac{9}{250}$); (d) $\frac{34}{25}$; (e) $\frac{63}{100}$;   (f) $\frac{33}{400}$

CHAPTER 2

Lesson 3

Writing Whole Numbers Using Numerals and Words

Writing Whole Numbers Using Numerals

Example 1   Write, using numerals: Forty-five million, seven hundred sixty-two thousand, three hundred twenty-five.

Note   1— One billion has nine zeros: 1,000,000,000
        2— One million has six zeros: 1,000,000
        3— One thousand has three zeros: 1,000
        4— One hundred has two zeros: 100
        5— One ten has one zero: 10

By applying these notes, let us write the sample number given above.

Step 1:   Forty-five million   = 45,000,000 (1 million has six zeros)
Step 2:   Seven hundred sixty-two thousand=   762,000 (1 thousand has three zeros)
Step 3:   Three hundred   =   300 (1 hundred has two zeros)
Step 4:   Twenty-five   =   25
Step 5:   Add the digits in the columns   _____.
                                        45,762,325

            Answer is: 45,762,325

Example 2   Write, using numerals: Fifty-four million, six hundred forty-five.

Step 1:   Fifty-four million: 54,000,000 (1 million has six zeros)
Step 2:   Six hundred: 600 (1 hundred has two zeros)
Step 3:   Forty-five: 45
Step 4:   Add the digits in the columns_____
                                    54,000,645

            Answer is: 54,000,645

Example 3   Write, using numerals: Two billion, eleven million, five thousand, seven.

        2,000,000,000   (1 billion has nine zeros)
           11,000,000   (1 million has six zeros)
                5,000   (1 thousand has three zeros)
                    7
        _____
        2,011,005,007

        Answer is: 2,011,005,007

Writing Whole Numbers in Words

In writing whole numbers in words, we will be guided by the group names:

billions millions thousands ones

$$\overset{\frown}{3},\overset{\frown}{296},\overset{\frown}{784},\overset{\frown}{625}.436954$$

Example 1 Write in words: 534

Solution We proceed as follows:

534 = 500 + 34 or 500 + 30 + 4

Five hundred thirty-four   (Hyphenate a number between 21 and 99. e.g., thirty-four)

Example 2 Write 12,379,534 in words
We proceed as follows (considering the digits in groups of three from right to left):
12,379,534

= 12,000,000 + 379,000 + 500 + 34

= Twelve million, three hundred seventy-nine thousand, five hundred thirty-four

Example 3 Write in words: 8,604,009
We proceed as follows:

8,604,009
= 8,000,000 + 604,000 + 9

= Eight million, six hundred four thousand, nine.

Lesson 3 Exercises

A. Write, using numerals: **1.** Five million, six hundred eight thousand, nine hundred seven.

2. Seven hundred twenty-nine thousand, four hundred.

3.. Thirty-six million, seventy thousand, two hundred thirty-four

Answers **1.** 5,608,907 ; **2.** 729, 400 ; **3.** 36,070,234

B. Write the following numbers using words:
1. 54,660,007; **2.** 735,038,429; **3.** 892,607

Answers **1..** Fifty-four million, six hundred sixty thousand, seven.
2. Seven hundred thirty-five million, thirty-eight thousand, four hundred twenty-nine.
3. Eight hundred ninety-two thousand, six hundred seven.

C. Write using numerals: **1.** Six billion, three hundred thousand, nine.

2. Eight hundred thirty-five thousand, nine hundred fifty-seven.

3. Two million, forty-seven thousand; **4.** One billion, ninety-five.
5. Four hundred thousand, twenty.

D. Write the following in words:

6. 834,689 ; **7.** 2,608,007; **8.** 629,353,086 ; **9.** 857.23 ; **10.** $487\frac{1}{2}$

Answers: **1.** 6,000,300,009; **2.** 835,957; **3.** 2,047,000; **4.** 1,000,000,095; **5.** 400,020
6. Eight hundred thirty-four thousand, six hundred eighty-nine; **7.** Two million, six hundred eight thousand, seven.
8. Six hundred twenty-nine million, three hundred fifty-three thousand, eighty-six;
9. Eight hundred fifty-seven and twenty-three hundredths; **10.** Four hundred eighty-seven and one half.

Lesson 4

Basic Arithmetic Operations and Properties

(The material in this lesson will guide us to justify the basic operations we perform in mathematics)

The four fundamental operations of arithmetic are addition (+), subtraction (-), multiplication (×) and division (÷). Each operation is an operation on two numbers. Two other operations are power finding (for example, $2^3 = 8$) and root extraction (for example, $\sqrt[3]{8} = 2$).

Addition

Example: $4 + 5 = 9$. The 4 and the 5 are called the **addends**, the 9 is called the **sum**.

Commutative Property: Addition is commutative. By this property, changing the order in which two numbers are added yields the same sum.
Example: $4 + 5 = 5 + 4$
$9 = 9$

Associative Property: Addition is associative. By this property, changing the grouping of numbers in addition yields the same sum.
Example: $(2 + 3) + 4 = 2 + (3 + 4)$.
$5 + 4 = 2 + 7$
$9 = 9$

Additive identity element The additive identity element is 0. For example, $5 + 0 = 5$

Identity property of 0. When zero is added to a given number, the sum is the given number.
Example: $4 + 0 = 4$.

Subtraction

Example: $7 - 2 = 5$. The 7 is called the **minuend**, the 2 is called the **subtrahend**, and the 5 is called the **difference**. Subtraction is the inverse of addition.
Subtraction is **not** commutative, That is, order is important in subtraction
For example, $6 - 2 \neq 2 - 6$ (That is, 6 - 2 is not equal to 2 - 6.)
$4 \neq -4$ (That is, 4 is not equal to -4.)
Subtraction is also **not** associative.
Example: $(8 - 5) - 2 \neq 8 - (5 - 2)$
$3 - 2 \neq 8 - 3$

Subtraction with Borrowing

Example   Subtract: $506 - 39$

Procedure: We use a vertical arrangement and indicate the decimal point
explicitly in all the whole numbers.
(Note that the explicit indication of the decimal point is **not** necessary)

Then, lining up the decimal points vertically, we obtain

$$\begin{array}{r} 506. \\ - \ 39. \\ \hline 467. \end{array}$$ (Check by adding 39 and 467 to obtain 506)

The answer is 467.

Note that, for a whole number, the decimal point is behind the units digit (last digit). The use of the decimal point, above, was to make sure that the "9" was under the "6" and the "3" was under the "0". (i.e., the units are under the units, and the tens are under the tens and so on.) Therefore, your may ignore the decimal points.

Multiplication

Implication of Multiplication Example: $2 \times 3 = 2(3) = (2)3 = (2)(3) = (2) \bullet (3) = 2 \bullet 3 = 6$
The 2 and the 3 are called the **factors,** and the 6 is called the **product.**

Commutative Property: Multiplication is commutative. By this property, changing the order in which two numbers are multiplied yields the same product.
Example: $2 \times 3 = 3 \times 2$
$6 = 6$

Associative Property: Multiplication is associative. By this property, changing the grouping of numbers in multiplication yields the same product.
Example: $(2 \times 3) \times 4 = 2 \times (3 \times 4)$.
$6 \times 4 = 2 \times 12$
$24 = 24$

Multiplicative identity: The multiplicative identity element is 1. For example, $6 \times 1 = 6$

Multiplicative inverse: The multiplicative inverse of a given number is that number which if multiplied by the given number yields a product equal to 1.

Example: The multiplicative inverse of 4 is $\frac{1}{4}$. (Note: $4 \times \frac{1}{4} = 1$.)

The multiplicative inverse of a number is also known as the **reciprocal** of that number.

Multiplicative property of zero Zero multiplied by any number equals zero.
Symbolically, for any whole number a, $a \times 0 = 0$.

Multiplication as repeated addition The consideration of multiplication as **repeated addition** can be a useful guide in solving problems such as the following: How many centimeters are there in 1000 meters? To solve this problem, we could add 100 centimeters a thousand times as:
$\underbrace{100 + 100 + 100 + 100 + ... + 100}_{\textbf{1000 addends}}$ (since 100 centimeters = 1 meter). However, instead of adding
100 a thousand times (a tedious computational process) we could simply multiply 100 by 1000 and obtain: $100 \times 1000 = 100{,}000$ centimeters .

Division

Implication of division   Example: $8 \div 2 = 4$ or $\frac{8}{2}$ or $2\overline{)8}^{\,4}$. The 8 is called the **dividend**, the 2 is called the **divisor**, and the 4 is called the **quotient**. Division is the inverse of multiplication. Note that $\frac{8}{2} = 4$ because $2 \times 4 = 8$.

Division is **not** commutative (That is, order matters in division.)
For example, $8 \div 2 \ne 2 \div 8$ (That is, $8 \div 2$ is **not** equal to $2 \div 8$)
$4 \ne \frac{1}{4}$ (That is, 4 is **not** equal to $\frac{1}{4}$.

Division is also **not** associative. For example,
$(18 \div 9) \div 3 \ne 18 \div (9 \div 3)$

$\frac{2}{3} \ne 6$

Division by 1: A given number divided by 1 equals the given number. For example, $\frac{4}{1} = 4$

Division involving zero

Examples 1. $\frac{0}{5} = 0$ (because $5 \times 0 = 0$)   **2.** $\frac{0}{-6} = 0$

   But   **3.** $\frac{5}{0}$ is undefined.(There is **no** number such that $0 \times$ "that number" $= 5$. Do not divide by 0)

       4. $\frac{0}{0}$ is indeterminate. (Any number will do since $0 \times$ "any number" $= 0$)

Division as repeated subtraction The consideration of division as **repeated subtraction** can be a useful guide in solving problems such as the following:
How many 60-gallon water containers can be filled from a tank containing 2,400 gallons of water? To solve this problem, we could subtract 60 gallons repeatedly from 2,400 gallons and add up the number of subtractions required for the tank to be either completely empty or to contain less than 60 gallons of water? However, instead of subtracting and recording the number of subtractions (40 subtractions), it is faster to divide 2,400 by 60 in one step to obtain 40 containers..

Powers of Whole Numbers

Examples 1. $2^5 = (2)(2)(2)(2)(2) = 32$

2^5 is read as 2 to the fifth power. (or 2 raised to the fifth power)
The exponent "5" indicates how many times the base "2" is being used as a factor.

2. $4^3 = 4 \times 4 \times 4$
     $= 64$
3. $10^4 = 10 \times 10 \times 10 \times 10$
       $= 10000$
4. $0^2 = 0 \times 0$
     $= 0$
5. $0^3 = 0 \times 0 \times 0$
     $= 0$

Zero as an Exponent

Any (nonzero) number raised the to power zero is 1.
Examples
     1. $5^0 = 1$
     2. $4^0 = 1$
     3. $10^0 = 1$

Note that 0^0 is indeterminate. Example: $0^0 = \frac{0^2}{0^2} = \frac{0}{0}$ which is indeterminate (see also division involving zero)

Roots of Whole Numbers

Root finding and **Power finding** are inverse operations (in much the same way as multiplication and division are inverse operations).

Square Root: Symbol " $\sqrt{}$ " or " $\sqrt[2]{}$ "

Finding the square roots of perfect squares by inspection or by factoring

1. In arithmetic, we may cheaply define the square root of a number as one of the two equal factors of that number. (In algebra, we shall modify this definition)

2. The square root of zero is zero (i.e. $\sqrt{0} = 0$).

Examples

1. The square root of 9 is symbolized $\sqrt{9}$, and

$$\sqrt{9} = 3$$

because $3^2 = 9$

Note that , according to the above definition,

$$\sqrt{9} = \sqrt{(3)(3)} = 3$$

2. The square root of 64 is written $\sqrt{64}$, and

$$\sqrt{64} = 8$$

because $8^2 = 64$.

Also, $\sqrt{64} = \sqrt{(8)(8)} = 8$

3. **Note:** $\sqrt{0} = 0$

because $0^2 = 0$

If the given number is not a perfect square, we will use tables or a calculator to find the square root.

The **Cube Root** : Symbol " $\sqrt[3]{}$ "

From the above square root definition, we can similarly define the cube root of a number as one of the three equal factors of that number. (In algebra, we shall modify this definition.)

Examples

1. $\sqrt[3]{8} = 2$

because $2^3 = 8$

Note that $\sqrt[3]{8} = \sqrt[3]{(2)(2)(2)} = 2$

2. $\sqrt[3]{64} = 4$

because $4^3 = 64$

Note that $= \sqrt[3]{4 \times 4 \times 4} = 4$

More formal definition of square root

1. $\sqrt{A} = r$ if $r^2 = A$, where A, and r are both real and positive.

2. The square root of zero is zero (i.e. $\sqrt{0} = 0$).

Lesson 4 Exercises

A. Subtract: **1.** 703 – 86 ;   **4.** 8,765,210 – 65,432

  2. 420 – 109 ;   **5.**   76,521 – 65,432

  3. 6035 – 247 ;   **6.**   894,352 – 743,956

Answers   **1.** 617;   **2.** 311 ;   **3.** 5,788; **4.** 8,699,778;   **5.** 11,089;   **6.** 150,396.

B. Evaluate the following:

  1. 3^3;   **2.** 2^4;   **3.** 8^2;   **4.** 4^0;   **5.** 0^0;   **6.** 0^3;   **7.** 4^3 ;   **8.** 2^4; **9.** 5^3

Answers: 1. 27;   **2.** 16 ;   **3.** 64 ;   **4.** 1 ;   **5.** Indeterminate ;   **6.** 0 ; **7.** 64 ; **8.** 16;   **9.** 125.

C. Simplify: **1.** $\sqrt{25}$;   **2.** $\sqrt{81}$;   **3.** $\sqrt{36}$; **4.** $\sqrt{256}$ **5.** $\sqrt{1024}$

  6. $\sqrt[3]{64}$ **7.** $\sqrt[4]{81}$;   **8.** $\sqrt{121}$;   **9.** $\sqrt{\frac{25}{49}}$;   **10.** $\sqrt[3]{8}$

Answers: **1.** 5 ;   **2.** 9 ;   **3.** 6 ;   **4.** 16 ;   **5.** 32 ;   **6.** 4 ;   **7.** 3 ;   **8.** 11 ;   **9.** $\frac{5}{7}$;   **10.** 2.

Word Problems

D

1. What must be added to 3 to make 7.?

2. What must be added to 86 to make 703.?

3. James wants to buy a computer costing $703. At present, he has only $86. How much more money does he need to have in order to buy this computer?

4. Mary had $806 in her bank account this morning. In the afternoon she withdrew $86 from this account to buy a printer, How much money is left in her bank account?

Answers: **1.** 4;   **2.** . 617;   **3.** . $ 617;   **4.** $ 617

Lesson 5

Order of Operations and
Evaluation of Arithmetic Expressions

Order of Operations in Performing Arithmetic Operations

Perform the operations according to the following order:

1. Grouping Symbols: If there are grouping symbols, evaluate within the grouping symbols first;
2. Then, evaluate the powers and the roots in any order ; followed by

3. Division and Multiplication from left to right (or invert the divisors and multiply in any order); and then

4. Addition and Subtraction from left to right.

Note: If grouping symbols occur within other grouping symbols, evaluate within the innermost grouping symbols first, followed by the next innermost symbols and so on.

Example 1

$5 + 2(4 + 7) - 9$
$= 5 + 2(11) - 9$ Evaluating within the parentheses first
$= 5 + 22 - 9$ Multiplying next
$= 27 - 9$ Adding from left to right
$= 18$

Example 2

$2^3 + 9(4 + 3) + 8$ Evaluating within the parentheses first

$= 2^3 + 9(7) + 8$
$= 8 + 63 + 8$ **Scrapwork:** $2^3 = (2)(2)(2) = 8$
$= 71 + 8$
$= 79$

Example 3

$18 \div 6 \times 3$ From left to right, we encounter division first and
$= 3 \times 3$ therefore, we divide first, according to
$= 9$ the order of operations.

Note in Example 3 above that, if we had multiplied first, we would **not** obtain the correct answer of 9.
In multiplying first, $18 \div 6 \times 3 = 18 \div 18 = 1$ <-----**This is the wrong solution.**

Example 4 $6 \times 3 \div 18$
$= 18 \div 18$ (From left to right, we encounter multiplication
$= 1$ first, therefore, we multiply first and then divide)
Note however that, if we divide first, we will still get the correct answer.

Thus, $6 \times 3 \div 18 = \dfrac{6}{1} \times \dfrac{3}{18} = \dfrac{6}{1} \times \dfrac{1}{6} = 1$

We conclude, from Examples 3 and 4, that if we always divide first (whether or not division comes first) , we will obtain the correct answer; but if multiplication does not come first in going from left to right, then we should **not** multiply first. Thus if we always perform division first, we will obtain the correct answer. Note that Example 3 and Example 4 are different problems.

EXTRA. Simplify: $20 \div \dfrac{1}{2}$ of 10 (Hint: Perform the "of" operation first before dividing.)

Solution $20 \div \dfrac{1}{2}$ of 10

     $= 20 \div (\dfrac{1}{2} \times 10)$ (The notion of changing "of" to multiplication and performing the operations from
     $= 20 \div 5 = 4$ left to right will lead to the **wrong result** in this problem: "of" precedes $\times$ and $\div$.)

Example 5
$$10 + 7[7 + 3(5 - 1)]$$
$$= 10 + 7[7 + 3(4)]$$
$$= 10 + 7[7 + 12]$$
$$= 10 + 7[19]$$
$$= 10 + [133]$$
$$= 143$$

Example 6 $\dfrac{18 + 6}{9 + 3} = \dfrac{24}{12}$
$$= 2$$

(Add the terms in the numerator, then
add the terms in the denominator
before dividing)

Example 7 Evaluate $3^2 + 7(3)$

Procedure: Apply the order of operations.
$$3^2 + 7(3)$$
$$= 9 + 21 \qquad \textbf{Scrapwork:} \quad 3^2 = 9$$
$$= 30$$

Example 8
$$3\sqrt{16} + 2(6) \div 3$$
$$= 3(4) + 12 \div 3$$
$$= 12 + 4$$
$$= 16$$

Example 9 Evaluate: 2×3^4

Solution : $= 2 \times 3 \times 3 \times 3 \times 3$
$$= 2 \times 81$$
$$= 162$$

Example 10 Evaluate: $2 + 4 \times (3 + 7)^2$

Solution: Apply the order of operations (see p.16)

$$2 + 4 \times (3 + 7)^2$$
$$= 2 + 4 \times (10)^2 \qquad \text{(Scrapwork } 3 + 7 = 10)$$
$$= 2 + 4 \times (100)$$
$$= 2 + 400$$
$$= 402 \qquad (\textbf{note} \text{ that the addition of the "2" was done last)}$$

Lesson 5 Exercises

A. **1.** $4 + 5 \times (6 + 5)$;   **2.** $8 \times (9 - 3) + 7^2$;   **3.** $38 + 7 \times (4 + 2)^2$

Answers   **1.** 59; **2.** 97; **3.** 290

B. Evaluate the following: **1.** $7 + 5 \times (3 + 4)$;   **2.** $8 + 3(5 + 1)^2$; **3.** $6\sqrt{81} + 5(4)(2)$

   4. $3[9 + (5 - 1) - 8]$;   **5.** $15 \div 3 \times 5$; **6.** $12 \times 8 \div 4$;   **7.** $\dfrac{4 + 7}{18 + 4}$;

8. $5 \times 2^3 + 7 \times 4$;   **9.** $2 \times 4 \times 6 \times 0 \times 9$;   **10.** $\dfrac{4 \times (3 - 1) - 8}{6}$; **11.** $\dfrac{5 + 2}{5 \times (3 - 2) - 5}$

Answers: **1.** 42; **2.** 116; **3.** 94; **4.** 15; **5.** 25; **6.** 24; **7.** $\dfrac{1}{2}$; **8.** 68; **9.** 0; **10.** 0; **11.** Undefined

C. Simplify the following:

   1. $6 + 2(3 + 8)$;   **2.** $12 \div 6 \times 2$;   **3.** $6 \times 2 \div 12$;   **4.** $\sqrt{4^2 + 3^2}$;

   5. $8^2 + 3(0)$;   **6.** $\dfrac{8 + 2}{2}$;   **7.** $\dfrac{6 + 3}{2 + 3}$;   **8.** $7 \times 6 + 2 \times 4$;

   9. $10^2 + 2 \times 4$;   **10.** $6\sqrt{9} + 2 \times 4$;   **11.** $3[4 + 2(3 + 1)] - 5$

Answers: **1.** 28; **2.** 4; **3.** 1; **4.** 5; **5.** 64; **6.** 5; **7.** $1\frac{4}{5}$; **8.** 50; **9.** 108; **10** 26; **11.** 31

Word Problems

D Last week, Paul had two part-time jobs. One job paid \$5 per hour, while the other job paid \$6 per hour.. He worked 5 hours last week at each job. In addition he was paid a \$4 bonus at one of the jobs.
What is the total amount of money Paul was paid, last week?

Answers: \$59

Lesson 6

Rounding off Whole Numbers and Estimation

Rounding off Whole numbers

Procedure:

Step 1: Locate the digit in the round-off place.
(The round-off place is the place to which we want to round-off the number)

Step 2: Drop all digits to the right of the round-off place, and if the digit immediately to the right of the round-off place is 5 or more, add 1 to the round-off place digit (i.e. we round-up); but if the digit immediately to the right of the round-off place is less than 5, the round-off place digit remains unchanged (i.e. we round-down). Also, replace each digit dropped by a zero.

Example 1 Round-off 85,376 to the following places:
(a) To the nearest ten.
(b) To the nearest hundred.
(c) To the nearest thousand.
(d) To the nearest ten-thousand.

Solution
(a) To the nearest ten

Step 1: The round-off place digit is 7 (since 7 is in the tens place)

Step 2: We drop the 6, add 1 to the 7 (since the 6 is more than 5) and replace the 6 by a zero.
Then 85,376
≈ 85,380   (Note the importance of replacing the 6 by zero. Without the zero, the answer
would be 8,538, which is far off the correct approximation of **85,380**.

Therefore 85,376 ≈ **85,380** (To the nearest ten)
(The symbol " ≈ " means " is approximately equal to".

(b)   85,376 to the nearest hundred

Step 1: The round-off place digit is 3. (since 3 is in the hundreds place)

Step 2: We drop the 7 and the 6, add 1 to the 3 (since 7 is more than 5) and replace the 7 and 6 by zeros.

Then 85,376
≈ 85,400

(c) 85,376 to the nearest thousand

Step 1: The round-off place digit is 5. (since 5 is in the thousands place)

Step 2: We drop the 3, the 7 and the 6; replace each of them by a zero, and retain the 5 unchanged
(since the 3 dropped is less than 5. The 3 is the digit immediately after the round-off place digit)

Then 85,376
≈ 85,000   (To the nearest thousand)

(d)   85,376 to the nearest ten-thousand

Step 1: The round-off place digit is 8.

Step 2: We drop the 5, the 3, the 7, and the 6; replace each of them by a zero and add 1 one to the 8.

Then 85,376 ≈ 90,000 (to the nearest ten-thousand)

Example 2   Round-off 73,964 to the nearest hundred.

Step 1: The round-off place digit is 9.(since 9 is in the hundreds place)

Step 2: We drop the 6, the 4; replace them by zeros, add 1 to the 9.
   (which becomes 10, and we write a zero to replace the 9 and add 1 to the 3 as usual)

Then 73,964
   $\approx$ 74,000 (To the nearest hundred)

Rounding off Decimals

Even though decimals are not be covered in this chapter (see Chapter 4), for some completeness, and comparison with rounding off whole numbers, we round-off some decimals, below.

We use the same rule for rounding off whole numbers, except that after the decimal point, we do **not** replace any digits dropped by zeros.

Example   We round off **85376.7463** to the following places:

   1. 85376.7463 to the nearest **thousandth** becomes **85376.746** (We do **not** replace the 3 dropped by a zero)

   2. 85376.7463 to nearest **hundredth** becomes **85376.75** (We added 1 to the digit in the round-off place)

   3. 85376.7463 to the nearest **tenth** becomes **85376.7** (The 7 is unchanged since the 4 dropped is less than 5)

   4. 85376.7463 the nearest **unit** becomes **85377.**   (Adding 1 to the 6)

   5. 85376.7463 to the nearest **ten** becomes **85380.** (Replacing the 6 dropped by a zero)

   6. 85376.7463 to the nearest **hundred** becomes **85400.** (Replacing the digits (6 and 7) dropped by zeros)

   7. 85376.7463 to the nearest **whole number** becomes **85377.** (same as to the nearest unit)

Estimation

In estimation, we **round-off** the numbers **before** carrying out the operation (which may be addition, subtraction, multiplication, division etc). We will round-off each number to the specified place; but if no round-off place is specified, we will round-off to the place of the first non-zero digit.
In the following example, we will round-off each number to the place of the first digit and carry out the indicated operation.

Example Estimate the product of 8,765 and 382.

Solution $9,000 \times 400$ (8,765 was rounded off to 9,000; and 382 to 400)
   $= 3,600,000$

Lesson 6 Exercises

A. Round-off **369,528** to the following places:
1. To the nearest ten.
2. To the nearest hundred.
3. To the nearest thousand.
4. To the nearest ten-thousand.
5. To the nearest hundred-thousand.

Answers: **1**. 369,530; **2**. 369,500; **3**. 370,000 ; **4**. 370,000 ; **5**. 400,000

B. Round-off **975.387** to the following places:

1. To the nearest hundredth.

2. To the nearest tenth.

3. To the nearest unit.

4. To the nearest ten.

5. To the nearest hundred.

6. To the nearest whole number.

Answers: **1**. 975.39 ; **2**. 975.4 ; **3**. 975; **4**. 980; **5**. 1,000 ; **6**. 975

C. Estimate the following: **1.**. $685,000 \times 364$; **2.** $6,457 + 346,238 + 7,398$;
3.. $8,672 \div 32$.

Answers: 1 . 280,000,000; **2** . 313,000 ; **3** . 300

Lesson 7

Factors, Product, Prime Numbers, Divisibility Rules, Prime Factorization

Factors and Product

Example   In $4 \times 9 = 36$, 4 and 9 are called the factors, and 36 is called the product.
(That is, the numbers being multiplied are called the factors, and the result of multiplying the numbers is called the product. Also, 4 and 9 are called divisors of 36. Therefore, saying that 4 is a factor of 36 implies that 4 divides 36 exactly (i.e., without a remainder or with zero remainder).

The process of finding the factors (divisors) of a given number is called **factoring.**

Prime Number: A natural number (greater than 1) whose natural number divisors are only itself and 1. (or a natural number (greater than 1) whose natural number factors are only itself and 1.)   Examples are 2, 3, 5, 7, 11, 13, 17, 19, 23,...

Composite Number: A natural number that has more natural number factors other than itself and 1.
Examples are 4, 6, and 10.
$$\big((4 = 2 \times 2 \times 1; \qquad 10 = 5 \times 2 \times 1)\big)$$

Divisibility Rules for 2, 3, and 5 (A guide for finding the factors 2, 3 and 5)

For 2: If the last (units) digit of a number is divisible by 2, then 2 is a divisor (or factor) of that number.
Examples: 2 is a divisor of the following:
5378; 6796; 310; 5552.

For 3: If the sum of the digits of a number is divisible by 3, then 3 is a divisor of that number.
Example: 3 is a divisor of **471**
Test: $4 + 7 + 1 = 12$. The sum 12 is divisible by 3.
The following are also divisible by 3 : **582; 7881; and 43263.**

The following are **not** divisible by 3: 673; 5431.

For 5: If the last digit of a number is 0 or 5, then 5 is a divisor (or factor) of that number.

Examples: 5 is a divisor of the following: **33420; 7775; 985; 111220.**

(See also next page for divisibility rules for 4, 6, 8 and 10.)

Prime Factor: A prime factor of a given number is a prime number which divides the given
number exactly.

Example 3 is a prime factor of 12 (because 3 is a prime number and it also divides 12 exactly).

However, 4 is **not** a prime factor of 12. (Although 4 is a factor of 12, it is not a prime number.)
Also, 5 is not a prime factor of 12. (Although 5 is a prime number, it is not factor of 12.)

Prime Factorization of a Number (Indicated product of all the prime factors of the number)

We will write a given number as the product of all the prime numbers which when multiplied produce the given number.

Example: Find the prime factorization of 84.

Solution: We will divide 84 successively by prime numbers which can divide it without remainders.

Step 1: We use 2 as divisor (since the last digit, 4, of 84 is exactly divisible by 2), followed by division by 3 and 7 (applying our knowledge of the multiplication tables), until the quotient is 1; and then, the division process is complete.

$$\begin{array}{r} 2\underline{|84} \\ 2\underline{|42} \\ 3\underline{|21} \\ 7\underline{|7} \\ 1 \end{array}$$

Step 2: The prime factorization of 84 is the product of the prime divisors used.

Answer: $84 = 2 \times 2 \times 3 \times 7$ or $2^2 \times 3 \times 7$

Another Method

Step 1: $84 = 7 \times 12$
Step 2: We factor the 12 into primes.
      $84 = 7 \times 2 \times 2 \times 3$
Answer: $84 = 2^2 \times 3 \times 7$.

Divisibility Rules for $4, 6, 8, 9,$ and 10 (See also previous page for the rules for 2, 3, and 5)

The divisibility rules for $4, 6, 8, 9$ and 10 are similar to the divisibility rules for $2, 3,$ and 5.

For 4: If the number named by the last **two digits** (from left to right) of a number is divisible by 4, then 4 is a divisor of that number. Example: 4 is a divisor of 1512 since 12 is divisible by 4.
(You may remember this rule as using the divisibility rule for 2 twice: Note that $2 \times 2 = 4$)

For 8: If the number named by the last **three digits** (from left to right) of a number is divisible by 8, then 8 is a divisor of that number. Example: 8 is a divisor of 2,1168 since 168 is divisible by 8.
(You may remember this rule as using the divisibility rule for 2 three times. Note that $2 \times 2 \times 2 = 8$)

For 6: If the sum of the digits of a number is divisible by 3 **and** the last digit is an even number, then 6 is a divisor of that number (This rule is a combination of the divisibility rules for 2 and 3).
Example: 6 is a divisor of 10584. (The sum of the digits, 18, is divisible by 3, and also the last digit is even)

For 9: A number is divisible by 9 if the sum of the digits is divisible by 9.
(You may remember this rule as using the divisibility rule for 3 twice. Note that $3 \times 3 = 9$)
Example: 12096 is divisible by 9 (because the sum $1 + 2 + 0 + 9 + 6 = 18$ is divisible by 9).

For 10. If the last digit of a number is 0, then 10 is a divisor of that number.
Example: 10 is a divisor of 3790.

Multiples and Common Multiples

Example   36 is a **multiple** of 4 because 36 can be obtained by multiplying 4 by a whole number, which in this case is 9. Similarly, 36 is a multiple of 9 because 36 can be obtained by multiplying 9 by a whole number, which in this case is 4. Also, 36 is divisible by 4 and by 9. Furthermore, 36 is a multiple of 12, because 36 can be obtained by multiplying 12 by a whole number (which in this case is 3). Also, 36 is divisible by 12.

Zero is a multiple of every whole number.

A **common multiple** is a number that is a multiple of two or more given whole numbers.

Also, a number (say, 36) is a common multiple of two or more given numbers (say, 9 and 12) if the number is divisible by each of the two or more given numbers (say, 9 and 12).
Thus, 36 is a common multiple of 9 and of 12, because 36 is divisible by 9 and by 12.

Definition: The **Least Common Multiple (LCM)** of two or more given (natural) numbers is the smallest (natural) number which is a multiple of the given numbers.
OR
The **Least Common Multiple (LCM)** of two or more given (natural) numbers is the smallest (natural) number which is divisible by each of the given numbers.

Finding the LCM (Least Common Multiple) of Two or More Numbers

Example Find the LCM of 16, 18, and 27.

Solution: We will cover two methods

Method 1

We will use a direct method which is the application of the prime factorization technique discussed previously. We will use **prime numbers** successively as divisors. At each step of the division process, each prime number used **must divide at least** one of the dividends (i.e., it may divide one, two or more, or even all the dividends), and the other dividends which are not divisible, are brought down for the subsequent steps. The division process ends when we have 1's in all the columns. The LCM is the product of the divisors used.

```
divisors
   |    dividends
   |    ↓    ↓    ↓
   ↓    ↓    ↓    ↓
  2 | 16 | 18 | 27
  2 |  8 |  9 | 27
  2 |  4 |  9 | 27
  2 |  2 |  9 | 27
  3 |  1 |  9 | 27
  3 |  1 |  3 |  9
  3 |  1 |  1 |  3
        1    1    1
```

Answer: The LCM of 16, 18 and $27 = 2 \times 2 \times 2 \times 2 \times 3 \times 3 \times 3$
$$= 432$$

Method 2 Finding the LCM by factoring

Step 1:   Factor each number into prime factors and write the repeated factors in power form.

$$16 = 2 \times 2 \times 2 \times 2 = 2^4$$
$$18 = 2 \times 3 \times 3 = 2 \times 3^2$$
$$27 = 3 \times 3 \times 3 = 3^3$$

Step 2: Consider every prime factor and choose the highest power of each prime factor.
The LCM is the product of the highest powers.

The LCM of 16, 18, and $27 = 2^4 \times 3^3 = 432$

Lesson 7 Exercises

A. Find the prime factorization of the following:

   1. 72;   **2.** 105;   **3.** 180 ;   **4.** 1024;   **5.** 432; **6.** 168 ; **7.** 420

Answers:

1. $2^3 \times 3^2$; **2.** $3 \times 5 \times 7$; **3.** $2^2 \times 3^2 \times 5$; **4.** 2^{10}; **5.** $2^4 \times 3^3$; **6.** $2^3 \times 3 \times 7$; **7.** $2^2 \times 3 \times 5 \times 7$

B. Find the LCM of the following: **1.** 9, 12 and 16;   **2.** 3, 7 and 8; **3.** 14, 15, and 20.

                                   4. 12, 18 and 32 ; **5.** 30 and 36;**6.** 15 and 24.

Answers **1.** 144 ; **2.** 168; **3.** 420; **4.** 288 ; **5.** 180; **6.** 120

CHAPTER 3
Fractions and Mixed Numbers

Lesson 8: **Definitions; Writing Fractions and Mixed Numbers in Words; Reducing Fractions to lowest terms; Equivalent Fractions**

Lesson 9: **Addition and Subtraction of Fractions**

Lesson 10: **Comparison of Fractions and Subtraction of Mixed Numbers**
Lesson 11: **Multiplication and Division: of Fractions and Mixed Numbers**

Lesson 8

Definitions; Writing Fractions and Mixed Numbers in Words; Reducing Fractions to lowest terms; Equivalent Fractions

Fraction: A fraction is a number representing one or more of the equal parts into which a whole (thing) has been divided.

Unit fraction One part of a whole which has been divided (or broken) into equal parts.

The numerator of a unit fraction is 1. Examples of unit fractions are $\frac{1}{2}, \frac{1}{2}$ and $\frac{1}{100}$.

Terms and Meaning of a Fraction

Example In the fraction, $\frac{2}{3}$, the 2 and 3 are called the terms of the fraction; the 2 is called the numerator and the 3 is called the denominator.

In a fraction, the numerator indicates how many of the unit fraction the fraction represents; the denominator indicates into how many equal parts a whole has been divided; the denominator also indicates the size of the unit fraction: the smaller the denominator, the larger the size, and he larger the denominator, the smaller the size of the unit fraction.

Note that in addition to representing parts of a whole, a **fraction** is also used to express **division**, and also to **compare** numbers.

Types of Fractions

Proper Fraction: A fraction in which the numerator is smaller than the denominator.

A proper fraction is always less than 1. Examples are $\frac{1}{2}, \frac{2}{7}$ and $\frac{34}{45}$

Improper Fraction: A fraction in which the numerator is greater than or equal to the denominator.

An improper fraction is greater than or equal to 1. Examples are $\frac{8}{5}, \frac{9}{3}$ and $\frac{7}{7}$.

Mixed Number: A mixed number consists of a whole number part and a proper fraction part. We obtain a mixed number from an improper fraction when the denominator does not divide the numerator exactly (i.e. without a remainder).

Converting an Improper Fraction to a Mixed Number

Example Convert $\frac{91}{17}$ to a mixed number.

Solution $\frac{91}{17}$ = 5 remainder 6 (using long division)

Therefore, $\frac{91}{17} = 5\frac{6}{17}$

$$\begin{array}{r} 5 \\ 17\overline{)91} \\ -85 \\ \hline 6 \end{array}$$

Converting a Mixed Number to an Improper fraction

Example Convert $6\frac{3}{4}$ to an improper fraction.

Step 1: $6 \times 4 = 24$
Step 2: $24 + 3 = 27$

Step 3: $6\frac{3}{4} = \frac{27}{4}$

(Normally, we do the conversion mentally and show only Step 3 (skipping Steps 1 & 2)).

Writing Fractions and Mixed Numbers in Words

Example 1   Write in words: $\frac{4}{5}$
Solution: Four fifths.

Example 2 Write in words: $7\frac{3}{10}$
Solution :   Seven **and** three tenths.

Example 3 Write in words: $568\frac{9}{100}$
Solution: Five hundred sixty-eight **and** nine hundredths.

Example 4 Write in words: $43\frac{2}{7}$
Solution   Forty-three **and** two sevenths.

Example 5: Write in words: $64\frac{102}{200}$
Solution: Sixty-four **and** one hundred two two-hundredths

A note about hyphenation

The author suggests that one should hyphenate only for clarity and to avoid confusion.. Unnecessary hyphenation sometimes leads to confusion in communication. The objective must be to communicate unambiguously.

Reduction of Fractions to Lowest terms

A fraction is in its lowest terms if the numerator and the denominator do not have any common factors (divisors) other than 1.

Examples of fractions in lowest terms: $\frac{2}{3}$, $\frac{5}{4}$ and $\frac{1}{6}$

Example Reduce $\frac{14}{21}$ to lowest terms.

Solution $\frac{14}{21}$

$\qquad = \frac{2 \times 7}{3 \times 7}$ <------You may skip this step.

$\qquad = \frac{2}{3}$ (Dividing out or canceling the common factor, 7.)

Equivalent Fractions

Equivalent fractions are fractions that have the same value but differ in the numerators and the denominators.

Examples: $\frac{1}{2}$, $\frac{2}{4}$, and $\frac{5}{10}$ are equivalent fractions. On reducing equivalent fractions to their lowest terms, they become identical. In the above examples, $\frac{1}{2}$ is in the lowest terms; $\frac{2}{4}$ and $\frac{5}{10}$ are higher terms of $\frac{1}{2}$

Forming Equivalent Fractions

Given a fraction **in** its lowest terms, to form an equivalent fraction, multiply both the numerator and the denominator by the same nonzero number.

Example 1 $\frac{1}{2} = \frac{1 \times 3}{2 \times 3} = \frac{3}{6}$

Given a fraction **not in** its lowest terms, to form an equivalent fraction, multiply (to higher terms) or divide (to lower terms) both the numerator and the denominator by the same nonzero number.

Example 2 $\frac{8}{12} = \frac{8 \times 2}{12 \times 2} = \frac{16}{24}$ <----------Going to higher terms.

Example 3 $\frac{8}{12} = \frac{8 \div 2}{12 \div 2} = \frac{4}{6} = \frac{2}{3}$ <-------Going to lower terms.

Example 4 Find the missing number: $\frac{8}{12} = \frac{?}{9}$

Solution

    Step 1: Reduce to lowest terms.

$$\frac{8}{12} = \frac{2}{3} = \frac{?}{9}$$

    Step 2: Form the required equivalent fraction.

$$\frac{2}{3} = \frac{2 \times 3}{3 \times 3} = \frac{6}{9}$$

The missing number is 6.

Note above (as in Example 4) that in some cases, if the given fraction is not in its lowest terms, first reduce to lowest terms before forming the equivalent fraction.

Example 5 Find x: $\frac{8}{12} = \frac{x}{9}$

Use exactly the same procedure as in Example 4. In the future, we will learn how to find x using algebraic methods.

Answer: $x = 6$.

Some applications of equivalent fraction formation

Equivalent fraction formation has a number of applications:

1. In adding and subtracting unlike fractions. (see p.30-31)

2. In forming proportions and solving proportion problems.(see p.59)

3. In rationalizing denominators of fractions involving radicals and complex numbers.

Lesson 8 Exercises

A. (a) Convert the following to mixed numbers: **1.** $\frac{22}{7}$; **2.** $\frac{17}{5}$; **3.** $\frac{79}{18}$.

(b) Verify the answers in **(a)** by converting your answers back to improper fractions.

(c) Change the following to improper fractions: **.1.** $8\frac{5}{6}$ **2.** $9\frac{3}{8}$

Answers: (a). **1.** $3\frac{1}{7}$; **2.** $3\frac{2}{5}$; **3.** $4\frac{7}{18}$; (c): **1.** $\frac{53}{6}$; **2.** $\frac{75}{8}$

B Write in words: **1.** $\frac{5}{8}$; **2.** $63\frac{4}{100}$; **3.** $2\frac{5}{17}$ **4.** $\frac{24}{25}$; **5.** $9\frac{3}{5}$; **6.** 63,400.

Answers: **1.** Five eighths; **2.** Sixty-three and four hundredths; **3.** Two and five seventeenths; **4.** Twenty-four twenty-fifths; **5.** Nine and three fifths; **6.** Sixty-three thousand, four hundred.

C. Write using numerals:: **1.** Eight and nine hundredths. **2.** Eight hundred nine. **3.** Eight hundred nine and nine tenths.

Answers: : **1.** $8\frac{9}{100}$; **2.** 809; **3.** $809\frac{9}{10}$

D. Reduce the following to lowest terms:

1. $\frac{21}{28}$ **2.** $\frac{24}{30}$ **3.** $\frac{20}{16}$ **4.** $\frac{512}{1024}$ **5.** $\frac{210}{1050}$

Answers: **1.** $\frac{3}{4}$; **2.** $\frac{4}{5}$; **3.** $\frac{5}{4}$); **4.** $\frac{1}{2}$; **5.** $\frac{1}{5}$

E **1.** Convert $\frac{3}{4}$ to 28ths (That is, change $\frac{3}{4}$) to an equivalent fraction whose denominator is 28.)

2. Convert $\frac{3}{2}$ to 16ths.

3. Change $\frac{5}{6}$ to 36ths

4. Find the missing number: $\frac{15}{16} = \frac{?}{32}$

5. Find x: $\frac{15}{16} = \frac{x}{32}$

6. Convert $\frac{11}{12}$ to 24ths.

7. Change $\frac{6}{12}$ to 20ths

8. Find the missing number: $\frac{?}{56} = \frac{5}{8}$

9. Find x: $\frac{x}{56} = \frac{5}{8}$

Answers: Answers: **1.** $\frac{21}{28}$; **2.** $\frac{24}{16}$; **3.** $\frac{30}{36}$; **4.** 30; **5.** 30; **6.** $\frac{22}{24}$); **7.** $\frac{10}{20}$; **8.** 35; **9.** 35.

Lesson 9 30

Addition and Subtraction of Fractions; Addition of Mixed Numbers

Addition of Fractions

Like Fractions (Fractions with the same denominator).

Example: Add $\frac{5}{8} + \frac{4}{8}$

Solution 1. Add the numerators and keep the common denominator.

2. Reduce the resulting fraction to lowest terms (if reducible) and change any improper fractions to mixed numbers.

$$\frac{5}{8} + \frac{4}{8} = \frac{9}{8}$$
$$= 1\frac{1}{8}$$

Unlike fractions (Fractions with different denominators).

Example 1 Add: $\frac{2}{7} + \frac{5}{9}$

In this case, we change the unlike fractions to like fractions, and then add the numerators

---

Format 1 (Sideways Format)	**Format 2 (Vertical Format)**
Step 1: Find the LCD of the fractions. (The LCD of the fractions is the LCM of the denominators. See p.24) The LCD of the fractions is 63.	$\frac{2}{7} + \frac{5}{9}$ Step 1: Find the LCD of the fractions The LCD = 63
Step 2: Change the fractions to equivalent fractions with LCD of 63. $$\frac{2}{7} = \frac{2 \times 9}{7 \times 9} = \frac{18}{63}$$ $$\frac{5}{9} = \frac{5 \times 7}{9 \times 7} = \frac{35}{63}$$	Step 2: Form equivalent fractions using the same LCD of 63 to obtain $$\frac{2 \times 9}{7 \times 9} + \frac{5 \times 7}{9 \times 7}$$ $$= \frac{18}{63} + \frac{35}{63}$$
Step 3: Add the numerators and keep the LCD, 63. $$\frac{18}{63} + \frac{35}{63} = \frac{53}{63}$$	Step 3: Add the numerators and keep the LCD, 63. to obtain $$\frac{53}{63}$$
Therefore, $\frac{2}{7} + \frac{5}{9} = \frac{53}{63}$.	Therefore, $\frac{2}{7} + \frac{5}{9} = \frac{53}{63}$.

Example 2 Add: $\frac{5}{6} + \frac{1}{2} + \frac{3}{4}$

Solution

Method 1

Example 2 Add: $\frac{5}{6} + \frac{1}{2} + \frac{3}{4}$

Format 1 (Sideways Format)

Step 1: Find the LCD of the fractions
  The LCD = 12

Step 2: Form equivalent fractions using
  the same LCD of 12.

Then $\frac{5 \times 2}{6 \times 2} = \frac{10}{12}$
  $\frac{1 \times 6}{2 \times 6} = \frac{6}{12}$
  $\frac{3 \times 3}{4 \times 3} = \frac{9}{12}$

Step 3: Add the numerators and keep
  the common denominator to obtain
  $$\frac{10 + 6 + 9}{12} = \frac{25}{12} = 2\frac{1}{12}$$

Format 2 (Vertical Format)
  $\frac{5}{6} + \frac{1}{2} + \frac{3}{4}$

Step 1: Find the LCD of the fractions
  The LCD = 12

Step 2: Form equivalent fractions using
  the same LCD of 12 to obtain

$= \frac{5 \times 2}{6 \times 2} + \frac{1 \times 6}{2 \times 6} + \frac{3 \times 3}{4 \times 3}$

$= \frac{10}{12} \quad + \quad \frac{6}{12} \quad + \quad \frac{9}{12}$

$= \frac{25}{12} = 2\frac{1}{12}$

Therefore $\frac{5}{6} + \frac{1}{2} + \frac{3}{4} = 2\frac{1}{12}$

**Method 2 Using prime factorization of the denominators followed by
  making the denominators of the fractions the same**

  Add: $\frac{5}{6} + \frac{1}{2} + \frac{3}{4}$

Step 1: Factor the denominators into primes.

  $\frac{5}{2 \times 3} + \frac{1}{2} + \frac{3}{2 \times 2}$

Step 2: Make the denominators have the same factors by looking for missing factors and
  multiplying accordingly; but note that whenever, you multiply the denominator by any
  factor, you must also multiply the numerator by the same factor (in order to keep the
  value of the original fraction unchanged).

  $\frac{5 \times 2}{2 \times 3 \times 2} + \frac{1 \times 3 \times 2}{2 \times 3 \times 2} + \frac{3 \times 3}{2 \times 2 \times 3}$ <---- The denominator of the first fraction needs a "2."
  The denominator of the second fraction needs
  a "2" and a "3".

  $= \frac{10}{12} \quad + \quad \frac{6}{12} \quad + \quad \frac{9}{12}$ The denominator of the third fraction needs a "3".

  $= \frac{10 + 6 + 9}{12}$

  $= \frac{25}{12}$

  $= 2\frac{1}{12}$

Subtraction of Fractions

The process in the subtraction of fractions is exactly like that of the addition of fractions, except
  that, in this case, we **subtract** the numerators (instead of adding) of the like fractions.

Addition of Mixed Numbers

Procedure: 1. Add the whole number parts.
　　　　　　2. Add the fractional parts and simplify.

Example Add: $2\frac{3}{4} + 3\frac{7}{8}$

Format 1 (Sideways Format)	**Format 2 (Vertical Format)**
$2\frac{3}{4} + 3\frac{7}{8}$	$2\frac{3}{4} + 3\frac{7}{8}$
$2\frac{3}{4} = 2\frac{6}{8}$	$= 2\frac{3\times 2}{4\times 2} + 3\frac{7}{8}$
$3\frac{7}{8} = 3\frac{7}{8}$	$= 2\frac{6}{8} + 3\frac{7}{8}$
	$= 5\frac{6+7}{8}$
$5\frac{13}{8} = 5 + 1\frac{5}{8}$　　Scrapwork	$= 5\frac{13}{8}$
$= 5 + 1\frac{5}{8}$　　$6 + 7 = 13$	$= 5 + 1\frac{5}{8}$
$= 6\frac{5}{8}$	$= 6\frac{5}{8}$

Lesson 9 Exercises

A Add the following: **1.** $\frac{5}{7} + \frac{4}{7}$;　**4.** $\frac{3}{10} + \frac{4}{15}$;　**7.** $\frac{2}{45} + \frac{5}{18} + \frac{7}{30}$;　**2.** $\frac{5}{12} + \frac{7}{9}$;

　　　　5. $\frac{3}{16} + \frac{5}{14} + \frac{7}{8}$;　**3.** $\frac{1}{6} + \frac{3}{5}$;　**6.** $\frac{8}{21} + \frac{1}{3} + \frac{9}{14}$

Answers 1. $1\frac{2}{7}$;　**2.** $1\frac{7}{36}$;　**3.** $\frac{23}{30}$;　**4.** $\frac{17}{30}$;　**5.** $1\frac{47}{112}$;　**6.** $1\frac{5}{14}$;　**7.** $\frac{5}{9}$

B Subtract: **1.** $\frac{9}{11} - \frac{5}{33}$;　**2.** $\frac{3}{7} - \frac{2}{7}$;　**3.** Subtract $\frac{1}{16}$ from $\frac{11}{48}$;　**4.** $\frac{7}{8} - \frac{5}{12}$;

　　　　5. $\frac{13}{18} - \frac{5}{12}$;　**6.** $\frac{5}{18} - \frac{2}{9}$;　**7.** $\frac{17}{48} - \frac{5}{16}$

Answers **1.** $\frac{2}{3}$;　**2.** $\frac{1}{7}$;　**3.** $\frac{1}{6}$;　**4.** $\frac{11}{24}$;　**5.** $\frac{11}{36}$;　**6.** $\frac{1}{18}$);　**7.** $\frac{1}{24}$

C Add: **1.** $5\frac{2}{3} + 7\frac{3}{5}$;　**3.** $6\frac{4}{5} + 7\frac{1}{5}$　　**5.** $4\frac{5}{8} + 7\frac{3}{4}$;　　**7.** $4\frac{8}{11} + 2\frac{1}{5}$

　　2. $4\frac{5}{12} + 9\frac{5}{8}$;　　**4.** $3\frac{2}{5} + 6\frac{7}{15}$;　**6.** $16\frac{3}{4} + 5\frac{1}{6}$;　　**8.** $2\frac{3}{4} + 3\frac{7}{8}$

Answers 1. $13\frac{4}{15}$;　**2.** $14\frac{1}{24}$;　**3.** 14;　**4.** $9\frac{13}{15}$　**5.** $12\frac{3}{8}$;　**6.** $21\frac{11}{12}$;　**7.** $6\frac{51}{55}$;　**8.** $6\frac{5}{8}$.

Lesson 10

Comparison of Fractions; Subtraction of Mixed Numbers

Comparison of Fractions

Example 1   Which fraction is the larger:

$$\frac{2}{5} \text{ or } \frac{4}{7} \text{ ?}$$

There are a number of methods for comparing fractions. We will use a method which we can call "cross-multiplication of the numerators and denominators". We will multiply the numerator of one fraction by the denominator of the other fraction. The fraction whose numerator (in the multiplication) results in the larger product is the larger of the two fractions; and the fraction whose numerator results in the smaller product is the smaller of the two fractions.

$$7 \times 2 = 14 \qquad 5 \times 4 = 20$$
$$\frac{2}{5} \diagup\!\!\!\!\diagdown \frac{4}{7}$$

(**Note** that the **directions** of the **arrows** are from the denominators to the numerators.)

Since 20 is larger than 14, $\frac{4}{7}$ is the larger of the two fractions (but of course, $\frac{2}{5}$ is the smaller of the two fractions). **Note** that $\frac{2}{5} = \frac{14}{35}$

and $\frac{4}{7} = \frac{20}{35}$

Example 2 Which is the larger:

$$\frac{7}{16} \text{ or } \frac{3}{5} \text{ ?}$$

$$5 \times 7 = 35 \qquad 16 \times 3 = 48$$
$$\frac{7}{16} \diagup\!\!\!\!\diagdown \frac{3}{5}$$

Since 48 is larger than 35, $\frac{3}{5}$ is the larger fraction (It is the fraction whose **numerator** when multiplied by the denominator of the other fraction resulted in the larger 48).

By using the above method , we do not have to find the least common denominator. We however, produce equivalent fractions, in which the common denominator may or may not be the least common denominator.

The objective of the above method was to find a fast method of comparing fractions.

Note: In some cases, if the LCD of the fractions can easily be found, we can form equivalent fractions using the same LCD, and then compare the numerators.

Determining the Smallest or Largest Fraction

Example 3 Determine the smallest fraction.

$$\frac{2}{3}, \ \frac{3}{5}, \ \frac{5}{7}, \ \frac{10}{13}$$

We will compare two fractions at a time, and after any comparison, we will keep the smaller fraction (since we want to determine the smallest fraction) and compare it with the other fractions.

Step 1:

$$5 \times 2 = 10 \qquad 3 \times 3 = 9$$

$$\frac{2}{3} \diagup\!\!\!\!\diagdown \boxed{\frac{3}{5}}$$

(We "box" the fraction to keep for the next step)

$\frac{3}{5}$ is the smaller (since 9 is smaller than 10) and we keep $\frac{3}{5}$ for the next step.

Step 2:

$$7 \times 3 = 21 \qquad 5 \times 5 = 25$$

$$\boxed{\frac{3}{5}} \diagup\!\!\!\!\diagdown \frac{5}{7}$$

(bringing down the next fraction, $\frac{5}{7}$, for comparison)

$\frac{3}{5}$ is the smaller (since 21 is smaller than 25)

Step 3:

$$13 \times 3 = 39 \qquad 5 \times 10 = 50$$

$$\boxed{\frac{3}{5}} \diagup\!\!\!\!\diagdown \frac{10}{13}$$

$\frac{3}{5}$ is the smaller (since 39 is smaller than 50)

The smallest fraction is $\frac{3}{5}$.

Note, above, that after every comparison, we 'box" the fraction to keep for the next comparison.

Example 4 For the last example, we will determine the largest fraction

$$\frac{2}{3}, \ \frac{3}{5}, \ \frac{5}{7}, \ \frac{10}{13}.$$

Bear in mind that in this example, we will always keep the larger fraction after every comparison (and compare it with the other fractions) since we want to determine the largest fraction.

Step 1: $5 \times 2 = 10$ $3 \times 3 = 9$

$$\boxed{\frac{2}{3}} \diagdown\!\!\!\!\diagup \frac{3}{5}$$

(We "box" the fraction to keep for the next step)

$\frac{2}{3}$ is the larger

Step 2: $7 \times 2 = 14$ $3 \times 5 = 15$

$$\frac{2}{3} \diagup\!\!\!\!\diagdown \boxed{\frac{5}{7}}$$

$\frac{5}{7}$ is the larger

Step 3: $13 \times 5 = 65$ $7 \times 10 = 70$

$$\frac{5}{7} \diagup\!\!\!\!\diagdown \boxed{\frac{10}{13}}$$

$\frac{10}{13}$ is the larger

The largest fraction is $\frac{10}{13}$

Arranging Fractions in Descending (decreasing) or Ascending (increasing) Order

To arrange in descending order: Determine the largest fraction first, remove it from the list and write it down; then determine the largest of the remaining fractions, and similarly remove it from the list and write it after the largest fraction. Repeat the process until only one fraction remains, and write this fraction last. (It is the smallest.)

To arrange in ascending order: The process here is similar to the descending order case, but in this case, we determine the smallest fraction first, followed by the next smallest and so on. The last remaining fraction becomes the largest, and it is written last.

Subtraction of Mixed Numbers

Case 1: We do **not** have to borrow

Example 1 Subtract $9\frac{5}{8} - 2\frac{1}{4}$

In this problem, because the fractional part being subtracted is smaller than the fractional part from which we are subtracting, we do **not** have to borrow from the whole number part; (Of course, if you borrow, you can still arrive at the correct answer, at the expense of extra simplification work).

$$9\frac{5}{8} = 9\frac{5}{8}$$
$$-2\frac{1}{4} = -2\frac{2}{8}$$
$$\overline{\phantom{-2\frac{1}{4} = } 7\frac{3}{8}}$$

Case 2: We have to borrow

Example 2   **Subtract:**
$$14\frac{3}{7} - 4\frac{5}{8}$$

A quick check (see p.33) indicates that $\frac{5}{8}$ is larger than $\frac{3}{7}$ ($8 \times 3 = 24$; $7 \times 5 = 35$)

Therefore we borrow 1 from the 14. (**The fractional part of the subtracted mixed number is larger than the fractional part of the mixed number from which we are subtracting.**)

Then, $14\frac{3}{7} = 13\frac{7}{7} + \frac{3}{7} = 13\frac{10}{7}$

Now, we subtract: $13\frac{10}{7} - 4\frac{5}{8}$

LCD = 56
$$13\frac{10}{7} = 13\frac{80}{56}$$
$$-4\frac{5}{8} = -4\frac{35}{56}$$
$$\overline{\phantom{-4\frac{5}{8} = } 9\frac{80-35}{56} = 9\frac{45}{56}}$$

Case 3: We have to borrow

Example 3 $12 - 4\frac{2}{5}$

$$\downarrow \qquad \downarrow$$

$$11\frac{5}{5} - 4\frac{2}{5} \quad \text{(we borrow 1 from the 12 and change it to } \tfrac{5}{5})$$

$$11\frac{5}{5}$$
$$-4\frac{2}{5}$$
$$\overline{\phantom{-4\frac{2}{5}} 7\frac{3}{5}}$$

(To check, add $4\frac{2}{5}$ and $7\frac{3}{5}$ to obtain 12)

Case 4: We do **not** have to borrow.   **Example 4**   Subtract $25\frac{4}{7} - 10$

$$25\frac{4}{7}$$
$$-10$$
$$\overline{ 15\frac{4}{7}}$$

In this case we do not have to borrow, since the number to be subtracted is a whole number (i.e., does not have a fractional part)

Lesson 10 Exercises

A. Determine the larger fraction:

1. $\frac{2}{7}$ or $\frac{4}{15}$; **2.** $\frac{5}{8}$ or $\frac{7}{11}$; **3.** $\frac{7}{8}$ or $\frac{8}{11}$; **4.** $\frac{3}{7}$ or $\frac{4}{9}$; **5.** $\frac{5}{6}$ or $\frac{9}{11}$

Answers **1**. $\frac{2}{7}$; **2.** $\frac{7}{11}$; **3.** $\frac{7}{8}$; **4.** $\frac{4}{9}$; **5.** $\frac{5}{6}$

B Determine the smallest fraction: **1.** $\frac{3}{5}$, $\frac{4}{7}$, $\frac{9}{13}$, $\frac{8}{11}$; **2.** $\frac{3}{4}$, $\frac{2}{3}$, $\frac{11}{15}$, $\frac{19}{30}$.

Answers **1.** $\frac{4}{7}$; **2.** $\frac{19}{30}$

C Determine the largest fraction:

1. $\frac{3}{5}$, $\frac{4}{7}$, $\frac{9}{13}$, $\frac{8}{11}$; **2.** $\frac{3}{4}$, $\frac{2}{3}$, $\frac{11}{15}$, $\frac{19}{30}$; **3.** $\frac{3}{8}$, $\frac{5}{16}$, $\frac{1}{4}$

Answers **1.** $\frac{8}{11}$; **2.** $\frac{3}{4}$; **3.** $\frac{3}{8}$

D Subtract the following: **1.** $6\frac{5}{12} - 2\frac{1}{4}$;   **5.** $3\frac{1}{4} - 1\frac{5}{8}$

2. $15\frac{2}{9} - 6\frac{5}{18}$;   **6.** $51\frac{4}{9} - 7\frac{1}{6}$

3. $48 - 7\frac{3}{8}$ **7.** $89 - 5\frac{3}{7}$

4. $76\frac{3}{4} - 23$ **8.** $43\frac{2}{5} - 15$

Answers **1.** $4\frac{1}{6}$; **2.** $8\frac{17}{18}$; **3.** $40\frac{5}{8}$; **4.** $53\frac{3}{4}$; **5.** $1\frac{5}{8}$; **6.** $44\frac{5}{18}$; **7.** $83\frac{4}{7}$; **8.** $28\frac{2}{5}$

E

Arrange the fractions in the above, B, and C in descending order; and in ascending order.

Word Problems

F

1. What must be added to 2 to make 5.?

2. What must be added to $2\frac{1}{4}$ to make $6\frac{5}{12}$.?

3. Maria needs $6\frac{5}{12}$ pounds of sugar to make some cakes, At present, she has only $2\frac{1}{4}$ pounds of sugar. How many more pounds of sugar does she need to have in order to make the cakes?

4. Mary had $6\frac{5}{12}$ pints of milk in her refrigerator this morning. In the afternoon she used $2\frac{1}{4}$ pints of this milk. How much milk is left in the refrigerator?

Answers: **1.** 3; **2.** $4\frac{1}{6}$ **3.** $4\frac{1}{6}$ pounds **4.** $4\frac{1}{6}$ pints

Lesson 11

Multiplication and Division: of Fractions and Mixed Numbers

Multiplication of Fractions

Procedure:

Step 1: Divide out (cancel) **any** common factors in the numerators and the denominators.

Step 2: Multiply the remaining numerators, and multiply the remaining denominators.

Example 1   Multiply $\frac{3}{5} \times \frac{5}{16} \times \frac{2}{9}$

Solution:

Step 1:   Canceling the common factors, we obtain

$$\frac{\overset{1}{\cancel{3}}}{\underset{1}{\cancel{5}}} \times \frac{\overset{1}{\cancel{5}}}{\underset{8}{\cancel{16}}} \times \frac{\overset{1}{\cancel{2}}}{\underset{3}{\cancel{9}}}$$

Step 2: Multiply the remaining factors.

$$\frac{1 \times 1 \times 1}{1 \times 8 \times 3}$$

$$= \frac{1}{24}$$

 Example 2   Multiply: $\frac{2}{3} \times \frac{2}{5}$

Solution: There are no common factors to cancel.

Therefore, we multiply the numerators, and then multiply the denominators.

$$\frac{2}{3} \times \frac{2}{5}$$

$$= \frac{2 \times 2}{3 \times 5}$$

$$= \frac{4}{15}$$

Division of Fractions

Reciprocals: The reciprocal of a real number A is $\frac{1}{A}$

 Examples: The reciprocal of $\frac{2}{3}$ is $\frac{3}{2}$. The reciprocal of 4 is $\frac{1}{4}$. The reciprocal of $\frac{1}{4}$ is 4. Thus, to find the reciprocal of a number, invert the number (or interchange the numerator and the denominator) The reciprocal of number is also known as the multiplicative inverse of that number.

The product of a number and its reciprocal is 1. Example $\frac{1}{4} \times \frac{4}{1} = 1$

Example 1 $\frac{4}{9} \div \frac{2}{15}$ ($\frac{4}{9}$ is the dividend; $\frac{2}{15}$ is the divisor)

Procedure: Invert the divisor and multiply. (or multiply the dividend by the reciprocal of the divisor)

$$\text{Then } \frac{4}{9} \div \frac{2}{15}$$

$$= \frac{4}{9} \times \frac{15}{2}$$

$$= \frac{\overset{2}{\cancel{4}}}{\cancel{9}} \times \frac{\overset{5}{\cancel{15}}}{\cancel{2}}$$
$$ \quad 3 \qquad 1$$

$$= \frac{2 \times 5}{3 \times 1}$$

$$= \frac{10}{3}$$

$$= 3\frac{1}{3}$$

(Always, cancel any common factors in the numerators and the denominators before multiplying)

Division of Complex Fractions

Example Simplify: $\dfrac{\frac{3}{4}}{\frac{16}{27}}$

Solution: $\dfrac{\frac{3}{4}}{\frac{16}{27}} = \frac{3}{4} \div \frac{16}{27}$

$$= \frac{3}{4} \times \frac{27}{16} \qquad \text{(inverting the divisor and multiplying)}$$

$$= \frac{81}{64}$$

$$= 1\frac{17}{64}$$

Multiplication of Mixed Numbers

Procedure: Step 1: Change each mixed number to an improper fraction.
Step 2: Multiply the resulting fractions.

Example $2\frac{4}{5} \times 2\frac{1}{7}$

Step 1: $= \frac{14}{5} \times \frac{15}{7}$

Step 2: $= \dfrac{\overset{2}{\cancel{14}}}{\cancel{5}} \times \dfrac{\overset{3}{\cancel{15}}}{\cancel{7}}$
$$ \quad 1 \qquad 1$$

$$= 6$$

Division of Mixed Numbers

Example $5\frac{2}{3} \div 4\frac{1}{2}$ ($4\frac{1}{2}$ is the divisor)

Procedure: Change each mixed number to an improper fraction and divide.

$$5\frac{2}{3} \div 4\frac{1}{2}$$

Step 1: $= \frac{17}{3} \div \frac{9}{2}$

Step 2: $= \frac{17}{3} \times \frac{2}{9} \qquad \text{(inverting the divisor and multiplying)}$

$$= \frac{34}{27} = 1\frac{7}{27}$$

Lesson 11 Exercises

A. Multiply:: **1.** $\frac{21}{32} \times \frac{8}{15}$; **2.** $\frac{3}{7} \times \frac{35}{36}$; **3.** $\frac{4}{5} \times \frac{6}{7}$;; **4.** $12 \times \frac{2}{3}$; **5.** $\frac{9}{8} \times \frac{4}{45}$

6. $\frac{3}{5} \times \frac{5}{16} \times \frac{2}{9}$

Answers **1**. $\frac{7}{20}$; **2.** $\frac{5}{12}$; **3.** $\frac{24}{35}$; **4.** 8 ; **5.** $\frac{1}{10}$; **6.** $\frac{1}{24}$

B. Divide: **1.** $\frac{4}{5} \div \frac{3}{7}$; **2.** $\frac{3}{8} \div \frac{5}{9}$; **3.** $\frac{5}{12} \div \frac{5}{4}$; **4.** $\frac{3}{4} \div \frac{4}{3}$; **5.** $\frac{7}{15} \div \frac{21}{20}$; **6.** $\frac{7}{8} \div \frac{3}{16}$

Answers: **1.** $1\frac{13}{15}$; **2.** $\frac{27}{40}$; **3.** $\frac{1}{3}$; **4.** $\frac{9}{16}$; **5.** $\frac{4}{9}$; **6.** $4\frac{2}{3}$

C. Multiply: (a) $3\frac{3}{4} \times 6\frac{2}{3}$; (b) $1\frac{2}{5} \times 1\frac{1}{2}$; (c) $1\frac{7}{9} \times \frac{3}{4}$

(d) $2\frac{1}{3} \times 5\frac{1}{4}$; (e) $3 \times 4\frac{1}{2}$; (f) $\frac{4}{5} \times 2\frac{1}{7}$; (g) $3\frac{3}{7} \times \frac{3}{4} \times 14$

Answers: (a) 25 ; (b) $2\frac{1}{10}$; (c) $1\frac{1}{3}$; (d) $12\frac{1}{4}$; (e) $13\frac{1}{2}$; (f) $1\frac{5}{7}$; (g) 36

D. Divide: (a) $5\frac{1}{2} \div 3\frac{1}{7}$; (d) $2\frac{2}{3} \div 5\frac{1}{3}$; (g) $7 \div 4\frac{2}{3}$

(b) $12\frac{3}{4} \div 2\frac{1}{8}$; (e) $1\frac{7}{8} \div 2\frac{3}{16}$; (h) $3\frac{1}{2} \div 1\frac{4}{5}$

(c) $14\frac{1}{4} \div 3$; (f) $2\frac{1}{8} \div 12\frac{3}{4}$

Answers (a) $1\frac{3}{4}$; (b) 6 ; (c) $4\frac{3}{4}$; (d) $\frac{1}{2}$; (e) $\frac{6}{7}$; (f) $\frac{1}{6}$; (g) $1\frac{1}{2}$; (h) $1\frac{17}{18}$

Test # 1   Student's Self-Test (**Always**, **Test yourself before you are tested**)   4 1
Attempt all questions on clean sheets of paper, **Do not write in the book** Show all necessary work.

---

1. Write, using numerals:

 Seven million, five hundred eight thousand, nine hundred eight.

2. Subtract: $8,765,210 - 64,434$

---

3. Round-off **965.467** to the following places:
(a). To the nearest hundredth.
(b) To the nearest tenth.

4. Write the following in words:
(a). 72,935,307;   (b) $586\frac{2}{3}$.

---

5. Estimate the following:
 (a) $674,000 \times 524$;
 (b) $7,457 + 356,238 + 7,393$.

6. Find the prime factorization of the following:
 (a)   160 ;   (b)   540.

---

7. Find the LCM of the following:
 (a) 18, 12 and 16;   (b) 2, 7 and 9.
(c) Find the smallest whole number that is divisible by each of 14, 42, and 56.

8. Convert the following to mixed numbers:
(a). $\frac{23}{7}$;   (b). $\frac{18}{5}$;   (c) $\frac{89}{18}$.

---

9. Change the following to improper fractions.
(a) $7\frac{5}{6}$;   (b) $8\frac{3}{8}$
Verify the answers in by converting your answers back to mixed numbers.

10. Write in words:
(a) $83\frac{4}{400}$;   (b) $\frac{34}{35}$;   (c) 63,400.

---

11. Write using numerals:
(a) Seven and three hundredths.
(b). Four hundred four and four tenths.

12. Reduce the following to lowest terms:
(a) $\frac{18}{30}$;   (b) $\frac{256}{512}$;   (c) $\frac{340}{1700}$

---

13.   Convert $\frac{3}{4}$ to 76ths

14. Convert $\frac{3}{12}$ to 28ths

15. Find x: $\dfrac{x}{24} = \dfrac{5}{8}$

16. Add: $\dfrac{7}{10} + \dfrac{3}{15}$

17. Add: $\dfrac{2}{15} + \dfrac{5}{18} + \dfrac{7}{30}$

18. Add: $6\dfrac{5}{12} + 9\dfrac{5}{8}$

19. Determine the smallest fraction:
$\dfrac{3}{7}, \quad \dfrac{4}{9}, \quad \dfrac{9}{13}, \quad \dfrac{8}{11}$

20. Subtract: $6\dfrac{5}{12} - 2\dfrac{1}{4}$

21. Subtract $98 - 6\dfrac{2}{9}$

22. Multiply: $\dfrac{7}{32} \times \dfrac{28}{15}$

23. Divide: $\dfrac{8}{15} \div \dfrac{21}{20}$

24. $4\dfrac{4}{7} \times \dfrac{3}{4} \times 14$

25. Divide: $14\dfrac{3}{4} \div 1\dfrac{1}{3}$

Bonus: Subtract: $86\dfrac{4}{7} - 38$

Answers: **1.** 7,508,908; **2.** 8,700,776; **3.** (a). 965.47, (b) 965.5;
4. (a) Seventy-two million, nine hundred thirty-five thousand, three hundred seven;
(b) Five hundred eighty-six and two thirds; **5.** (a) 350,000,000; (b) 414,000;
6. (a) $2^5 \times 5$; (b) $2^2 \times 3^3 \times 5$; **7.** (a) 144; (b) 126; (c) 168; **8.** (a) $3\frac{2}{7}$; (b) $3\frac{3}{5}$; (c) $4\frac{17}{18}$;
9. (a) $\frac{47}{6}$; (b) $\frac{67}{8}$; **10.** (a) Eighty-three and four four-hundredths; (b) Thirty-four thirty-fifths;
(c) Sixty-three thousand, four hundred; **11.** (a) $7\frac{3}{100}$ (b) $404\frac{4}{10}$;
12 (a) $\frac{3}{5}$; (b) $\frac{1}{2}$; (c) $\frac{1}{5}$; **13.** $\frac{57}{76}$; **14.** $\frac{7}{28}$; **15.** 15 ; **16.** $\frac{9}{10}$; **17.** $\frac{29}{45}$; **18.** $16\frac{1}{24}$;
19. $\frac{3}{7}$; **20.** $4\frac{1}{6}$; **21.** $91\frac{7}{9}$; **22.** $\frac{49}{120}$; **23.** $\frac{32}{63}$; **24.** 48 ; **25.** $11\frac{1}{16}$; **Bonus:** $48\frac{4}{7}$

CHAPTER 4
DECIMALS

Lesson 12: **Definitions, Addition and Subtraction; Multiplication and Division by Powers of 10, Multiplication of Decimals; Comparison of Decimals ; converting decimals to fractions**

Lesson 13: **Complex Decimals; Dividing Decimals; Converting Fractions to Decimals**

Lesson 12

Definitions, Addition and Subtraction of Decimals; Multiplication and Division by Powers of 10; Multiplication of Decimals; Comparison of Decimals ; Converting Decimals to Fractions

Definitions

Powers of 10

Examples:

$$10^1 = 10$$
$$10^2 = 100$$
$$10^3 = 1,000$$
$$10^4 = 10,000$$

A **decimal fraction** is a fraction whose denominator is a power of 10.

Examples are $\frac{3}{10}$, $\frac{5}{100}$, and $\frac{23}{1000}$..

A **decimal** is the symbol for an equivalent decimal fraction.

Examples: $\frac{3}{10} = .3$ ($\frac{3}{10}$ is the decimal fraction; .3 is the decimal)

$$\frac{5}{100} = .05$$

$$\frac{23}{10000} = .0023$$

Some Common Fractions, Decimal Fractions, Decimals and Names

Common Fraction	Decimal Fraction	Decimal	Name
$\frac{1}{2}$	$\frac{5}{10}$	.5	Five tenths
$\frac{1}{4}$	$\frac{25}{100}$	.25	Twenty-five hundredths
$\frac{61}{100}$	$\frac{61}{100}$	.61	Sixty-one hundredths
$\frac{490}{1000}$	$\frac{490}{1000}$	.490	Four hundred ninety thousandths

Number of decimal places or decimal digits: The number of decimal places is the number of digits to the right of the decimal point.

Examples: (a) 0.62 has two decimal places; (b) 46.3 has one decimal place; (c) 53.005 has three decimal places.

Also note for example that $0.62 = .62$ (The zero preceding the decimal point on the left-hand side is used to remind the reader that a decimal point follows.)

Equivalent decimals

Equivalent decimals are decimals that have the same value but differ in the number of decimal places.

(see also p. 28 for a similar definition of equivalent fractions)

Example: .5 , .50 , and .500 are equivalent decimals, but note that .5 is the simplest form

(in much the same way as $\frac{1}{2}$ is the lowest terms of the equivalent fractions $\frac{2}{4}$ and $\frac{5}{10}$).

Similarly, 4.2 and 4.20 and 4.2000 are equivalent decimals

Addition of Decimals

Example   Add: $13.23 + 7.1 + 9$

Procedure: Indicate the decimal point explicitly in all the whole numbers.
Arrange the numbers vertically, while lining up the decimal points.
Note: $9 = 9$.

Then we obtain:

```
        ——— decimal points in line
        ↓
    13.23   (Add as is done in whole
     7.10    number addition)
     9.00
    29.33
```

Subtraction of Decimals

Example: Subtract: $15 - 2.43$

Indicating the decimal point in the 15 and arranging the numbers vertically, while lining up the decimal points, we obtain:

```
    15.00<------------------------(attach zeros)
    −2.43
    12.57
```

Multiplying decimals or whole numbers by a power of 10

Procedure: Move the decimal point to the right as many places as there are zeros in the power of 10.

Examples of powers of 10

Example 1 $3.45 \times 10 = 34.5$ $10^1 = 10$

Example 2 $3.45 \times 10^2 = 345.$ $10^2 = 100$

Example 3 $3.45 \times 10^3 = 3450.$ $10^3 = 1000$

Example 4 $23 \times 10000 = 230000.$ or 230000
(In Examples 3 and 4, we wrote zeros to hold places)

Dividing decimals or whole numbers by a power of 10

Procedure: Move the decimal to the left as many places as there are zeros in the power of 10.

Example 1 $53.4 \div 10 = 5.34$

Example 2 $53.4 \div 100 = .534$

Example 3 $53.4 \div 1000 = .0534$

Example 4 $534 \div 10 = 53.4$

Example 5 $\frac{672.89}{100} = 6.7289$

Example 6 $\frac{89675}{100} = 896.75$

Multiplication of a decimal by a decimal or by a whole number

Step 1: Ignore the decimal points and multiply the whole numbers.

Step 2: Count the total number of decimal places (decimal digits) in the decimals being multiplied.

Step 3: Insert the decimal point in the product from Step 1, so that the number of decimal places equals the total number of decimal places counted in Step 2. (We may have to write zeros to hold places in some cases.)

Example Multiply 4.21 by 3.5   **Note**: $4.21 \times 3.5 = \frac{421}{100} \times \frac{35}{10} = \frac{14735}{1000} = 14.735$

$$
\begin{array}{r}
4.2\,1 \\
\times\ 3.5\ \underline{} \\
\hline
2105 \\
1263 \\
\hline
14.735
\end{array}
$$

(The total number of decimal places is 3)

Comparison of Decimals

Example   Determine which decimal is the smallest: .234, .098, .0725

There are a number of methods for comparing decimals.
First, we will use a method which we can call " comparison of place-value digits", and then we will discuss another method.

Method 1

Step 1:   Arrange the numbers vertically while lining up the decimal points **as if** one were to add the numbers.

.2340   (we can attach zeros)
..0980
.0725

Step 2:   We will compare the digits from left to right (tenths, hundredths, and thousandths places etc).
In the tenths' column, 2 is the largest, and therefore .2340 cannot be the smallest. We cross out .2340 or ignore it.

Step 3: Compare   .0980
.0725

Since the tenths' places are the same, we go to the next column (the hundredths' place) and there, 7 is smaller than 9 and therefore .
.0725 is smaller than .098 . Hence .0725 is the smallest decimal.
(of course, .2340 is the largest decimal)

Method 2

Multiply each decimal by a power of 10, so that all the decimals become whole numbers.
 In multiplying by a power of 10, move the decimal point to the right as many places as there are zeros in the power of 10. The power of 10 to use is based on the decimal with the most number of decimal places.

In the above example, .0725 has the most number of decimal places, four decimal places.

Proceeding,

$$.2340 \times 10000 = 2340$$
$$..0980 \times 10000 = 980$$
$$.0725 \times 10000 = 725$$

Since .0725 gave the smallest whole number 725, .0725 is the smallest decimal.

Arranging Decimals in Descending (decreasing) or Ascending (increasing) Order

To arrange in descending order: Determine the largest decimal first, remove it from the list and write it down; then determine the largest of the remaining decimals, and similarly remove it from the list and write it after the largest decimal. Repeat the process until only one decimal remains, and write this decimal last. (It is the smallest decimal.)

To arrange in ascending order: The process here is similar to the descending order case, above, but in this case, we determine the smallest decimal first, write it down, followed by the next smallest, and so on. The last remaining decimal becomes the largest, and it is written last.

Converting a Decimal to a Fraction

Example   Change .96 to a fraction in its lowest terms.

Step 1: Write .96 as a decimal fraction .

$$.96 = \frac{96}{100}$$

(The number of zeros in the denominator equals the number of decimal places in the decimal)

Step 2:   Reduce the fraction to its lowest terms.

Then, $\dfrac{\cancel{96}}{\cancel{100}} = \dfrac{24}{25}$

$$\therefore \ .96 = \frac{24}{25}$$

Lesson 12 Exercises

A. Write as decimals: (a) $\frac{36}{1000}$; (b) $\frac{40}{100}$; (c) $\frac{9}{100}$

Answers (a) .036 ; (b) .40 ; (c) .0 9

B. Add the following:: (a) $13.456 + 4.03 + 9 + .463$; (d) $19 + .018 + 298$
(b) $43.050 + 73.4 + 7 + .0304 + 29$; (e) $.5 + .54 + .6407$
(c) $7.8 + 47 + 5.605 + .006$; (f) What must be added to 3.48 to make 9?.

Answers (a). 26.949; (b) 152.4804 ; (c) 60.411 ; (d) 317.018 ; (e) 1.6807 ; (f) 5.52

C. Subtract: (a) $60.54 - 3.679$; (b) $28 - 16.603$; (c) $46.07 - 9$; (d) $8.23 - 5.025$
(e) Subtract 37.5 from 76.4; (f) From 8, subtract 5.96; (g) Subtract .65 from 1.1.

Answers (a) 56.861; (b) 11.397 ; (c) 37.07; (d) 3.205; (e) 38.9 ; (f) 2.04 ; (g) .45

D. Multiply: (a) 65.8×100; (b) 4.306×10^2; (c) $.006 \times 10000$; (d) 26×1000

Answers (a) 6580. or 6580; (b) 430.6 ; (c) 60. or 60; (d) 26,000.

E. Divide:(a) $74.68 \div 100$; (b) $34.07 \div 10^4$; (c) $68953 \div 100$; (d) $\frac{245.6}{10000}$

Answers (a) .7468 ; (b) .003407 ; (c) 689.53 (d) .02456

F. Which decimal is the larger? (a) .0764 or .04685 (b) 3.245 or .986753
(c) 4 or .8954

Answers (a) .0764 ; (b) 3.245 ; (c) 4

G. Determine the smallest decimal: (a) .067, .0096, .12402; (b) 3.008, .094, .4509
(c) 1.007, .753, .087

Answers (a) .0096; (b) .094; (c) .087

H. Use the inequality symbol " > " to compare the following:
(a) .0764 or .04685; (b) 3.245 or .986753; (c) 4 or .8954

I. Order the following using the symbol " < ": (a) .067, .0096, .12402;
b) 3.008, .094, .4509; (c) 1.007, .753, .087

J. 1. Arrange the decimals in **G** and **H** (above) in descending order..
2. Arrange the decimals in **G** and **H** (above) in ascending order..

K. Change each of the following to a fraction in its lowest terms:
(a) .64 ; (b) 6.4 ; (c) 0.036 ; (d) 1.36 ; (e) .63; (f) 0.0825

Answers (a) $\frac{16}{25}$) ; (b) $\frac{32}{5}$; (c) $\frac{9}{250}$; (d) $\frac{34}{25}$; (e) $\frac{63}{100}$; (f) $\frac{33}{400}$

Lesson 13

Complex Decimals; Dividing Decimals; Converting Fractions to Decimals

A complex decimal consists of a decimal part and a common fraction part, Examples are $.11\frac{5}{7}$ and $.117\frac{1}{7}$.

Using complex decimals allows one to write the exact quotient of a division problem in which no round-off place for the quotient is specified. Example: $.82$ divided by $7 = .11\frac{5}{7}$ or $.117\frac{1}{7}$

In a complex decimal, the **number of decimal places** equals the number of digits to the right of the decimal point, **ignoring** the fractional part.

Examples   **1.** $.8\frac{2}{3}$ has **one** decimal place.   **2.** $.14\frac{1}{3}$ has **two** decimal places.

   3. $.3\frac{1}{3}$ has **one** decimal place.   **4.** $.33\frac{1}{3}$ has **two** decimal places.

   5. $.166\frac{2}{3}$ has **three** decimal places.

Converting a Complex Decimal to a Fraction

Example 1   Convert $.8\frac{2}{3}$ to a fraction in its lowest terms.

$$.8\frac{2}{3} = \frac{8\frac{2}{3}}{10} \qquad (.8\frac{2}{3} \text{ has one decimal place; and note, for example, that } .8 = \frac{8}{10})$$

$$= \frac{26}{30} \qquad \left(\frac{26}{3} \div \frac{10}{1} = \frac{26}{3} \times \frac{1}{10} = \frac{26}{30} \right)$$

$$= \frac{13}{15}$$

Example 2   Convert $.14\frac{1}{3}$ to a fraction in its lowest terms.

$$.14\frac{1}{3} = \frac{14\frac{1}{3}}{100} \qquad (.14\frac{1}{3} \text{ has two decimal places})$$

$$= \frac{43}{300}$$

How to divide a Decimal by a Whole number

Procedure Align the decimal point vertically in the quotient and divide as in whole number division. You may attach zeros (one at a time) if needed to continue the division process.

Case 1: Round-off place for the quotient is specified

Example 1 Divide .82 by 7 and round-off quotient to the nearest hundredth.

$$.117 \approx .12 \qquad \text{(We divided to the thousandths place before rounding-off)}$$

$$
7\overline{).82}
$$

$$
\begin{array}{r}
-7 \\
\hline
12 \\
-7 \\
\hline
50 \\
-49 \\
\hline
1
\end{array}
$$

Conclusion: To the nearest hundredth, $\dfrac{.82}{7} \approx .12$

Case 2: No Round-off place for the quotient is specified

In this case, we will leave the answer as a complex decimal, since this will be equivalent to the exact quotient. Note that when we round-off, the quotient obtained is an approximation.

Example 2 Divide .82 by 7. (Round-off-place is **not** specified)

$.11\frac{5}{7}$ (dividing to **two** decimal places) $.117\frac{1}{7}$ ((dividing to **three** decimal places)

$$
7\overline{).82} \qquad\qquad\qquad 7\overline{).820}
$$

$$
\begin{array}{r}
-7 \\
\hline
12 \\
-7 \\
\hline
5
\end{array}
\qquad \text{or} \qquad
\begin{array}{r}
-7 \\
\hline
12 \\
-7 \\
\hline
50 \\
-49 \\
\hline
1
\end{array}
$$

To check: Let us convert $.11\frac{5}{7}$ and $.117\frac{1}{7}$ back to the original division problem.

$$.11\tfrac{5}{7} = \frac{82}{700} \qquad\qquad .117\tfrac{1}{7} = \frac{820}{7000} = \frac{82}{700}$$

$$= \frac{.82}{7} \qquad\qquad\qquad\quad = \frac{.82}{7}$$

Therefore, $.11\frac{5}{7}$ and $.117\frac{1}{7}$ are equivalent.

We will make the following agreement: In computations, unless otherwise instructed, we will leave the decimal as a complex decimal of two decimal places. This will produce sufficient decimal places for most applications and yet provide additional information for either obtaining more decimal places or for rounding-off.

Lesson 13: Complex Decimals; Dividing Decimals; Converting Fractions to Decimals

To **round-off** a complex decimal by **dropping only** the **fractional part,** add 1 to the digit immediately to the left of the fractional part if the numerator of the fractional part is greater than or equal to half of the denominator; but if the numerator is less than half of the denominator, the digit to the left of the fractional part remains unchanged.

Examples (a): $.11\frac{5}{7}$ rounded-off to the nearest hundredth becomes .12 $\qquad \left(5 > \frac{7}{2}\right)$

(b) $.117\frac{1}{7}$ rounded-off to the nearest thousandth becomes .117 $\qquad \left(1 < \frac{7}{2}\right)$

Note however that $.117\frac{1}{7}$ rounded-off to the nearest hundredth becomes .12

Dividing a Number by a Decimal
Procedure

Step 1: Make the divisor a whole number by moving the decimal point to the right as many places as the number of decimal places in the divisor, and at the same time, move the decimal point in the dividend to the right as many places as the decimal point was moved in the divisor. Note however, that for a divisor such as .20 or .200, we need to move the point only one place to the right to make it a whole number, since 2 is equivalent to 2.0 or 2.00.

Step 2: Rewrite the dividend and the divisor and divide as by a whole number.

Example 1 Divide 1.23 by .8 and round-off quotient to the nearest hundredth.

Step 1: $.8\overline{)1.23}$ = $.8\overline{)1.23}$ (moving the decimal point 1 place to the right in the divisor and dividend)

Step 2: $8\overline{)12.3}$

Step 3:
$$
\begin{array}{r}
1.537 \approx \mathbf{1.54} \\
8\overline{)12.300} \\
\underline{-8} \\
43 \\
\underline{-40} \\
30 \\
\underline{-24} \\
60 \\
\underline{-56} \\
4
\end{array}
$$

Conclusion: To the nearest hundredth, $\dfrac{1.23}{.8} \approx \mathbf{1.54}$

Example 2 Divide .86 by .20

Step 1: $.20\overline{)\,.86}$ = $.20\overline{)\,.86}$ (We need to move the decimal point only one place to make it a whole number)

Step 2: $2\overline{)8.6}$ with quotient 4.3

Writing Fractions as Decimals 5 1

When a fraction (a rational number) is converted to a decimal, the decimal is either terminating or repeating

Examples of terminating decimals: **1.** $\frac{1}{4}$ = .25;   **2.** $\frac{1}{8}$ = .125

Examples of repeating decimals: **1.** $\frac{1}{3}$ = .$\overline{3}$ or .33...;   **2.** $\frac{3}{11}$ = .$\overline{27}$ or .2727...

Converting a Fraction to a Decimal

Case 1: Round-off place of the decimal is specified

Example   Change $\frac{3}{7}$ to a decimal, rounding off the quotient (answer) to the nearest
          hundredth (two decimal places).

Procedure: Using long division, divide the numerator by the denominator, attaching zeros to the dividend during
          the division process. We will carry the division to three decimal places (that is, one extra place)
          before rounding off the quotient. Do **not** just stop at the hundredth place.

$$
\begin{array}{r}
0.428 \\
7\overline{)3.000} \\
-28 \\
\hline
20 \\
-14 \\
\hline
60 \\
-56 \\
\hline
4
\end{array}
$$

Then, 0.428 to the nearest hundredth becomes 0.43.

Note: To round off quotient to the nearest tenth, we carry out the division to two decimal places before
rounding off;   to the nearest thousandth, the division is to four decimal places.

Case 2: Round-off place of the decimal is **not** specified

In this case, we will leave the decimal as a complex decimal of two decimal places, since this will produce
 sufficient decimal places for most applications and yet provide additional information for either obtaining
more decimal places or for rounding-off.

Example 2   Change $\frac{3}{7}$ to a decimal.   (Round-off place is **not** specified)

Procedure: Using long division, divide the numerator by the denominator, attaching zeros to the dividend during
          the division process. No round-off place is specified and therefore we will carry the division to two d
          ecimal places, and write the remainder divided by the divisor as a fraction.

$$
\begin{array}{r}
0.42\frac{6}{7} \quad \longleftarrow \text{(complex decimal with two decimal places)} \\
7\overline{)3.000} \\
-28 \\
\hline
20 \\
-14 \\
\hline
6
\end{array}
$$

Thus $\frac{3}{7}$ = $0.42\frac{6}{7}$ (a complex decimal with two decimal places)

Example 3: Change $\frac{2}{3}$ to a decimal rounded-off to two decimal places.

$$
\begin{array}{r}
0.666 \approx .67 \\
3\,\overline{)\,2.000} \\
-\underline{1\,8} \\
20 \\
-\underline{18} \\
20 \\
-\underline{18} \\
2
\end{array}
$$

Example 4: Change $\frac{2}{3}$ to a decimal.

$$
\begin{array}{r}
0.66\frac{2}{3} \longleftarrow \text{(complex decimal with two decimal places)} \\
3\,\overline{)\,2.00} \\
-\underline{1\,8} \\
20 \\
-\underline{18} \\
2
\end{array}
$$

Note: In Example 4, we could write the quotient as $0.\overline{6}$ or $0.6...$ instead of $0.66\frac{2}{3}$. However, in computations, it is much easier to convert $0.66\frac{2}{3}$ to a fraction than to convert $0.\overline{6}$ or $0.6...$ to a fraction. The general method for converting $0.\overline{6}$ or $0.6...$ to a fraction is algebraic.

Lesson 13 Exercises

A. Determine the number of decimal places

1. $.2\frac{2}{3}$;   2. $.02\frac{2}{3}$;   3. $1.2\frac{1}{3}$;   4. $.4\frac{3}{11}$;   5. $43.4\frac{2}{7}$

Answers: **1.** 1 ;   **2.** 2;   **3.** 1 ;   **4.** 1 ;   **5.** 1 .

B. Convert to a fraction in its lowest terms.

1. $.2\frac{2}{3}$;   2. $.02\frac{2}{3}$;   3. $1.2\frac{1}{3}$;   4. $.4\frac{3}{11}$;   5. $43.4\frac{2}{7}$

Answers: **1.** $\frac{4}{15}$;   **2.** $\frac{2}{75}$;   **3.** $\frac{37}{30}$;   **4.** $\frac{47}{110}$;   **5.** $\frac{304}{7}$

C. Divide: **1.** $0.72 \div 5$; **2.** $4.32 \div 12$; **3.** $72.1 \div 0.20$; **4.** $0.738 \div 0.6$; **5.** $30.03 \div 2.1$

Divide and round-off quotient to the nearest hundredth.

6. $0.463 \div 0.8$; **7.** $62.82 \div 14$; **8.** $56.6536 \div 23$; **9.** $46.95 \div 1.1$; **10.** $38.93 \div 0.08$

Answers: **1.** 0.144;   **2.** 0.36;   **3.** 360.5;   4, 1.23;   **5.** 14.3;   **6.** 0.58;   **7.** 4.49;   **8.** 2.46;
   9. 42.68; **10.** 486.63

D. **1.** Change $\frac{6}{7}$ to a decimal, rounding off the quotient to the nearest hundredth.

   2. Change $\frac{5}{9}$ to a decimal, rounding off quotient to the nearest tenth.

Answers **1.** $0.857 \approx 0.86$ **2.** $0.55 \approx 0.6$

CHAPTER 5
Ratio

Lesson 14: **Definition; Reduction of Ratios to Lowest Terms**
Lesson 15: **Using Ratios to Compare Quantities**
Lesson 16: **Using Ratios to Divide a Quantity into Parts**

Note: This chapter and the next chapter are very important for everyday life, whether you have your own business or you work for someone. If you master these chapters very well, you would have a good mathematical preparation for a lot of endeavors both at home or at the workplace., We have used both arithmetic and algebraic methods in covering these chapters. If you have no algebraic background, you may skip the algebraic parts and cover only the arithmetic parts. However, you can study the algebraic material in chapters 9-13 before covering these chapters.

Lesson 14

Definition and Reduction of Ratios to Lowest Terms

Whenever you go shopping to buy apples, you always want to know the selling price, and the first question you usually ask the seller is "How much are these apples?" A typical answer may be "4 apples for one dollar". As soon as you have agreed to this price, you and the seller have established a ratio between the number of apples the seller is willing to give you and the number of dollars you are willing to pay. Simply, the ratio you have agreed to is 4 apples to 1 dollar, and we can symbolize this as 4:1. The seller might also have replied "1 dollar for 4 apples", but in this case, the ratio of dollars to apples is 1 to 4 or 1:4. Note the order in which the terms 1 and 4 of the ratio follow "dollar and apples".

If your grandmother left $1000, in her will, to be divided between you and your brother so that you receive $600 and your brother receives $400, then the ratio of your share to your brother's share is $600 to $400 and in simplest terms, this ratio is **3 to** 2, which means that whenever you receive $3, your brother receives $2.

If the simple interest rate for your savings account at a bank is 5% (5 percent or 5 per hundred), then after one year, the ratio of the interest to the deposit you had in your account is 5 to 100 (That is, your interest is $5 for every $100 in your account.)

We can observe from the above examples that, in everyday life, we are always dealing with ratios, consciously or unconsciously. Uses of ratios include comparison of quantities, division of a quantity into parts, in business decision making and in forming proportions.

Definition
The **ratio** of two quantities a and b or of a to b (written $a{:}b$) is a comparison of the two quantities and written as a fraction is $\frac{a}{b}$ (i.e., is the first quantity mentioned divided by the second quantity mentioned), where a and b may or may not be of the same units. The quantities a and b are called the terms of the ratio. However, the ratio of b to a is $\frac{b}{a}$.

Examples

1. The ratio, 3**:**4 is $\frac{3}{4}$ but the ratio 4**:**3 is $\frac{4}{3}$. **Note** that order is important.

2. The ratio, 4 inches to 12 inches is the fraction $\frac{4\ \text{inches}}{12\ \text{inches}} = \frac{1}{3}$
   (Terms have the same units and cancel out)

3. The ratio, 4 inches to 1 foot is the fraction $\dfrac{4 \text{ inches}}{1 \text{ foot}} = \dfrac{4 \text{ inches}}{12 \text{ inches}} = \dfrac{1}{3}$ (1 ft = 12 in.)

(One unit can be converted to the other)

4. The ratio, 4 apples to 1 dollar is the fraction $\dfrac{4 \text{ apples}}{1 \text{ dollar}}$

(Terms do not have the same units and one unit cannot be converted to the other)

5. The ratio, 50 miles to 2 hours is the fraction $\dfrac{50 \text{ miles}}{2 \text{ hours}} = \dfrac{25 \text{ miles}}{1 \text{ hour}}$ (25 mph) <-different units

Note that since ratios can be written as fractions, ratios can be reduced to lowest terms in the same way that we reduce fractions to lowest terms.

Since ratios are used for comparing quantities, a ratio may have three or more terms.

Example The number of pencils, the number of pens, and the number of notebooks are in the ratio of 2:3:4. (i.e., 2 pencils to 3 pens to 4 notebooks)

Rate

If a and b are **not** of the same units **and** one unit **cannot** be converted to the other unit, some authors call the ratio a **rate.**

1. The rate, 50 miles to 1 hour is the fraction $\dfrac{50 \text{ miles}}{1 \text{ hour}}$ (50 mph)

Even though **rate** is a common terminology for quantities with different units, in the following example, note that it is actually a ratio, since the terms have the same units or no units.

2. An interest rate of 5% for a savings account (at a bank in the US.) means that the depositor receives $5 for every $100 in the account. Note also that 5% is also the fraction $\dfrac{5}{100}$ or the ratio, 5:100, which is a ratio. Perhaps we should call this interest rate interest ratio.

Example: Reduce the ratio 12:8 to lowest terms

Solution: $12:8 = \dfrac{12}{8} = \dfrac{3}{2}$

Answer: In lowest terms 12:8 = 3:2.

Lesson 14 Exercises

Reducing Ratios to Lowest Terms

Reduce to lowest terms (Two terms)
(a). The ratio 2:4
(b). The ratio 6:3
(c). The ratio 8:12
(d) The ratio 12:8
(e). The ratio 5 inches to 10 inches
(f) The ratio 6 apples to 3 dollars

Answers: (a) 1:2; (b) 2:1; (c) 2:3; (d) 3:2; (e) 1:2; (f) 2 apples: 1 dollar;

Reduce to lowest terms (Three Terms)
(a) .The ratio 2:4:6   (b) The ratio 6:3:18
(c) The ratio 8:12:6
(d) The ratio 12:8:6
(e) . The ratio 5 inches to 10 inches to 30 inches
(f) The ratio 6 apples to 10 apples to 24 apples
Answers: (a) 1:2:3; (b) 2:1:6; (c) 4:6:3; (d) 6:4:3; (e) 1:2:6; (f) 3:5:12;

Lesson 15

Using Ratios to Compare Quantities

Example 4

At a certain college, there are 2,000 male students and 6,000 female students.

(a) What is the ratio of the number of male students to the number of female students at this college?

(b) What is the ratio of number of female students to the number of male students at this college?

(c) What is the ratio of the number of female students to the total number of students at this college?

Solution

(a) The ratio of males to females is $2000{:}6000 = \frac{2000}{6000} = \frac{1}{3}$ or $1:3$

(b) The ratio of females to males is $6000{:}2000 = \frac{6000}{2000} = \frac{3}{1}$ or $3{:}1$

(c) Total number of students $= 2000 + 6000 = 8000$
The ratio of the number of females to the total number of students is $6000{:}8000$
$= \frac{6000}{8000} = \frac{3}{4}$ or $3:4$

Example 5

Now let us verify the results of the above problem by doing it backwards.

There are altogether 8,0000 students at a certain college. The ratio of the number of males to the number of females at this college is 1:3.

(a) Find the number of males.   **(b)** Find the number of females at this college.

Solution

(a) Fraction of students who are males $\frac{1}{1+3} = \frac{1}{4}$ (Fraction corresponds to first term, 1, of ratio)

The number of males $= \frac{1}{4} \times \frac{8000}{1} = 2{,}000$

(b) Fraction of students who are females $\frac{3}{1+3} = \frac{3}{4}$ (Fraction corresponds to second term, 3)

The number of females $= \frac{3}{4} \times \frac{8000}{1} = 6{,}000$

(c) Let us recalculate the number of females given that the ratio of number of females to the total number of students is $3{:}4$.

Solution: Fraction of students who are females $= \frac{3}{4}$.

The number of females $= \frac{3}{4} \times \frac{8000}{1} = 6{,}000$ <-----This is the same solution as is (b)

Note the difference between how the denominator, 4, of the fraction in (c) was obtained by using only the second term of the ratio, and how the denominator of the fraction in (b) was obtained by adding 1 and 3, the terms of ratio. Note also that in (b) and (a), we added the terms of the ratio to obtain the denominators of the fractions, since the terms, 1 and 3 correspond to the individual genders involved in the problem; while in (c), the second term, 4, of the ratio corresponds to the total number of students. Note particularly that we did **not** add the 3 and the 4 to obtain the denominator of the fraction in (c), but we rather used only the 4 as the denominator. See also p.73. The problem in (c) could have been posed as " 3 out of 4 students at a certain college are females. If there are 8,000 students at this college, how many of the students are females. The solution method is the same as in (b). We could also algebraically, use proportion to obtain the same solution: $\frac{3}{4} = \frac{x}{8000}$, where x is the

number of females.

$4x = 3 \times 8000$ and $x = \frac{3}{4} \times \frac{8000}{1} = 6,000$.

As we shall soon observe, Example 5 actually belongs to the next lesson.

Lesson 15 Exercises

Comparing two quantities

1. At a certain college, there are 18,000 male students and 32,000 female students.

  (a) What is the ratio of the number of male students to the number of female students at this college?

(b) What is the ratio of number of female students to the number of male students at this college?

 c) What is the ratio of the number of female students to the total number of students at this college? **Answer** (a) 9:16; (b) 16:9 ; (c) 16:25

2. In a math class, class, there are 15 females and 8 males.

  (a) What is the ratio of the number of males to the number of females in this class?

  (b) What is the ratio of the number of females to the number of males?

  (c) What is the ratio of the number of males to the total number of students?

                          Answer: (a) 8:15; (b) 15:8; (c) 8:23

3. Rameltha and Albert are to divide $45,000

    If Rameltha receives $18,000, what is the ratio of Albert's share to Rameltha's share?

                        Answer: 3:2

4. A piece of black tape is 4 inches long, and a similar piece of red tape is 2 feet long..

    What is the ratio of the length of the black tape to the length of the red tape?

                        Answer: 1:6

5. Out of 20 games that a team played, the team won 12 games and lost the rest.

  (a) What is the ratio of the number of games won to the number of games played?

  (b) What is the ratio of the number of games won to the number of games lost?

                      Answer: (a) 3:5; (b) 3:2

Comparing three quantities

6. In a bag, there are 16 red marbles, 20 black marbles and 12 green marbles.

  (a) What is the ratio of red marbles to black marbles to green marbles?

  (b) What is the ratio of black marbles to red marbles to green marbles?

  (c) What is the ratio of black marbles to the sum of the red and green marbles?

  (d) What is the ratio of green marbles to the total number of marbles in the bag?

      Answer: (a) red:black:green = 4:5:3 ; (b) black:red:green = 5:4:3; (c) 5:7; (d) 1:4

7. Rameltha, Albert and James are to divide $45,000. If Rameltha receives $18,000, and Albert receives $17,000, what is the ratio of Rameltha's share to James' share to Albert's share?   Answer: 18:10:17

8. A piece of black tape is 4 inches long, and similar pieces of red and green tapes are

    2 feet , and 3 feet long respectively.

    (a) What is the ratio of the length of the black tape to green tape to red tape?

    (b) What is the ratio of the length of green tape to red tape to black tape?

    Answer: (a) 1:9:6   (b) 9:6:1

9. Out of 20 games that a team played, the team won 12 games, lost 4 and drew the rest

  (a) What is the ratio of games won to games lost to games drawn?

  (b) What is the ratio of games lost to games won to games played?

  (c) What is the ratio of games drawn to games won?

              Answer: (a) 3:1:1 ; (b) 1:3:5 ; (c) 1:3

Lesson 16

Using Ratios to Divide a Quantity into Parts

Example 1   Divide 28 in the ratio 3: 4

Method 1:   Using Arithmetic

Step 1:   Fraction of the number corresponding to the first term: $\frac{3}{3+4} = \frac{3}{7}$

Fraction of the number corresponding to the second term: $\frac{4}{3+4} = \frac{4}{7}$

Step 2: The number corresponding to the first term (of the ratio) $= \frac{3}{7} \times \frac{28}{1} = 12$

The number corresponding to the second term (of the ratio) $= \frac{4}{7} \times \frac{28}{1} = 16$

The numbers are 12 and 16.

Method 2: Using algebra

Let the number corresponding to the first term (of the ratio) $= 3k$ *(k* a constant)
Let the number corresponding to the second term (of the ratio) $= 4k$
"The sum of the two numbers is 28" translates to
$$3k + 4k = 28$$
$$7k = 28$$
$$k = 4$$
One of the numbers $= 3k = 3(4) = 12$, and
the other number $= 4k = 4(4) = 16$

The numbers are 12 and 16.

Method 3:   Using Algebra

See Method 3 of Example 2, below.

Example 2 A bag contains red and white marbles in the ratio 3:4. There are altogether 28 marbles
in this bag. How many of these marbles are red?

Solution Method 1:   Using arithmetic

The number of red marbles $= \frac{3}{7} \times \frac{28}{1} = 12$

Method 2: Using algebra

Let the number of red marbles $= 3k$ *(k* a constant)
Let the number of white marbles $= 4k$
$$3k + 4k = 28$$
$$7k = 28 \quad \text{The number of red marbles} = 3k = 3(4) = 12$$
$$k = 4$$

Method 3: Using Algebra

Let the number of red marbles $= x$
Then the number of white marbles $= (28 - x)$
The proportion "3:4 as $x : (28 - x)$ translates to: (ratios of red marbles to white marbles)
$$\frac{3}{4} = \frac{x}{(28 - x)}$$

Solve for x:
$$3(28 - x) = 4x$$
$$84 - 3x = 4x$$
$$84 = 7x$$
$$12 = x$$
The number of red marbles $= 12$

Lesson 16 Exercises

Dividing a quantity into two parts

1. The ratio of male students to female students at a certain college is 2:5. if there are 2,800 students at this college, how many of the students are females? Answers: **1.** 2,000;

2. If the ratio (by weight) of hydrogen to oxygen in water is 1:8, find the weight of hydrogen in 27 grams of water.   Answer: 3 grams

3. The sum of two numbers is 40. The numbers are in the ratio 2:3. Find these numbers.
                              Answers: **1.** 16 and 24

4. James and Mary are to divide $72 in the ratio 5:7. How much does each receive?
                              Answer: James: $30, Mary: $42

5. A grandmother left $45,000 in her will to be divided between two granddaughters, Maria and Diana in the ratio 2 to 3. How much does each receive?
                              Answer Maria: $18,000; Diana: $27,000

6. John invested $2,000 in a company while James invested $3,000 in the same company. If after 3 years, a profit of $2,000 is to be shared between them in the ratio of their investments. How much does each receive?   **Answer** John: $800; James: $1200

7. Mary wants to prepare a solution consisting of alcohol and water in the ratio 1:2 by volume. If the total volume of the solution is 12 liters, find the volumes of alcohol and water to be used
                              Answer alcohol: 4 liters; water: 8 liters

9. A farmer wants to prepare 24 liters of gasohol which consists of alcohol and gasoline in the ratio 1:9. How many liters of alcohol and gasoline are to be mixed?
                              Answer alcohol: 2.4 liters; gasoline: 21.6 liters

9. A bag contains p fruits, made up of oranges and apples only in the ratio q:r. Obtain an expression for the number of oranges in terms of p, q, and r.Answer: $\frac{pq}{q+r}$

10. In a math class, the ratio of males to females is 1:3.: If there are 28 students in this class, how many males are in this class? Answer: 7 males

Dividing a quantity into three parts

11. A grandmother left $45,000 in her will to be divided between three granddaughters, Ana, Elizabeth, and Diana in the ratio 1 to 2 to 3. How much does each receive?
                    Answer: Ana: $7,500, Elizabeth: $15,000, Diana: $22,500

12. John invested $3,000 in a company while James and Alexander invested $4,000 and $5,000 respectively in the same company. If after 4 years, a profit of $4,800 is to be shared between them in the ratio of their investments. How much does each receive?
Answer : John: $1,200, James: 1,600, Alexander: $2,000

13. Mary wants to prepare a solution consisting of alcohol, water and orange juice in the ratio 1:**2**:**3** by volume. If the total volume of the solution is 36 liters, find the volumes of alcohol, water and juice to be used. **Answer** Alcohol: 6 liters, water: 12 liters, orange juice: 18 liters

14. A farmer wants to prepare 44 liters of fuel blend which consists of fuel A, fuel B and fuel C in the ratio 1:1:9. How many liters of each type of fuel are to be mixed?
                    Answer : fuel A: 4 liters, : fuel B: 4 liters, fuel C: 36 liters

15. A bag contains p fruits, made up of oranges, grapefruits, and apples in the ratio q:r:s.
Obtain an expression for the number of oranges in terms of p, q, r, and s. Answer: $\frac{pq}{q+r+s}$

16. On a math test, the number of A's, B's, and C's are in the ratio of 1:4:2 If there are 28 students in this class, determine the number of (a) A's, (b) B's, and (c) C's. ?
                    Answer: A's: 4, B's: 16, C''s: 8:

CHAPTER 6

PROPORTION (VARIATION)

Lesson 17

Introduction to Direct and Inverse Proportion

(Definitions, Terminology, Types of Proportion and Principles)

Basically, there are two main **types** of **proportion** (or **variation**):
1. **Direct proportion** or simply, **proportion** (or direct variation).
2. **Inverse proportion** (or indirect variation).
   There are also problems which involve both direct and inverse proportion simultaneously.

Direct proportion
Discussion: In the introduction of the lesson on ratios, we discussed an example involving buying apples. There, the seller quoted the price of apples as "4 apples for 1 dollar", and we established the ratio of the number of apples to the number of dollars as 4 to 1. Now, if the buyer wants to buy exactly 4 apples, then the buyer pays exactly 1 dollar. However, if the number of apples is different from 4, then the number of dollars to pay would be different from 1, and the seller has to determine how much the buyer pays for the new number of apples. For example, if the buyer wants 8 apples (instead of 4 apples), then the seller will determine how many dollars the buyer pays (in this case, $2), or if the buyer wants 2 dollars worth of apples, the seller has to determine how many apples to give the buyer for $2 (in this case, 8 apples). We note above that if the buyer wants to buy exactly 4 apples, then only two quantities are involved in the problem, namely 4 apples and 1 dollar.

However, if the buyer wants 8 apples, then four quantities would be involved in the problem, namely 4 apples, 1 dollar, 8 apples, and the price for 8 apples. We will determine the fourth quantity, namely, the price for 8 apples using these three known quantities and a relationship called a **direct proportion**. In this relationship, as the number of apples increases, the number of dollars also increases; or as the number of apples decreases, the number of dollars also decreases. When the number of apples is zero, the number of dollars is also zero. Uniform or constant changes in the number of apples result in uniform changes in the number of dollars. For example, as the number of apples is doubled, the number of dollars is also doubled; but if he number of apples is halved, the number of dollars is also halved.
   Another application of proportion is as follows: Suppose the roof of your house has a triangular shape, and you want to build a much bigger or smaller house such that the corresponding roof has the same shape as your present house, then **direct proportion** can be used to determine the lengths of the sides of the new roof.

Direct proportion occurs very often in the arithmetic of everyday life. It may not be the number of apples and number of dollars, but it may be the number of miles traveled and number of gallons of gasoline used, or the number of miles traveled and number of hours, or the number of grams of hydrogen and the number of grams of oxygen. Direct proportion is usually termed **proportion** in most textbooks, and it is this type that you meet very often.

We should note above that problems stated in terms of proportion can also be stated in terms of variation and conversely. In the examples covered in this chapter, we cover both cases.

Simple Direct Proportion (Direct Variation) 60

Definition: A **proportion** is a statement that two ratios are equal.
The direct proportion "a is to b as c is to d " (or $a:b::c:d$) is the equality

$$\frac{a}{b} = \frac{c}{d} \quad \text{(quotient = quotient)}$$

In the above proportion, a and d are called the extremes and b and c are called the means of the proportion. We may also define a proportion as a mathematical statement indicating the equality between two **equivalent fractions.** Example: $\frac{2}{3} = \frac{4}{6}$ is a proportion

Proportion Principles

1. We can invert both ratios of a proportion: From (1) above, we obtain after the inversion,

$$\frac{b}{a} = \frac{d}{c}$$

2. We can interchange the extremes of a proportion: From (1) above we would obtain:

$$\frac{d}{b} = \frac{c}{a}$$

3. We can also interchange the means to obtain:

$$\frac{a}{c} = \frac{b}{d}$$

Also $ad = bc$ (That is the product of the extremes = the product of the means) <-- also called cross-multiplication

The proportion principles can be used to obtain desired ratios. For example, if we have $\frac{3}{2}$ as the ratio of one side of a proportion, and we desire the ratio $\frac{2}{3}$ then proportion principle #1 above (inversion of both sides of a proportion) can be used to obtain this ratio, $\frac{2}{3}$, from $\frac{3}{2}$.

More Principles (From $\frac{a}{b} = \frac{c}{d}$, we obtain the following:

4. $\frac{a+b}{b} = \frac{c+d}{d}$ (by addition). Note: $\frac{a+b}{b} = \frac{a}{b} + \frac{b}{b} = \frac{a}{b} + 1$ **and** $\frac{c+d}{d} = \frac{c}{d} + \frac{d}{d} = \frac{c}{d} + 1$

5. $\frac{a-b}{b} = \frac{c-d}{d}$ (by subtraction)

6. $\frac{a-b}{a+b} = \frac{c-d}{c+d}$ (by subtraction and addition)

7. $\frac{a+b}{a-b} = \frac{c+d}{c-d}$ (by addition and subtraction)

Note:: For 6 and 7, divide LHS and RHS of 4 and 5,

8. If $\frac{a}{b} = \frac{c}{d} = \frac{e}{f}$, then $\frac{a+c+e}{b+d+f} = \frac{a}{b} = \frac{c}{d} = \frac{e}{f}$

e.g.,($\frac{2}{10} = \frac{3}{15} = \frac{4}{20} = \frac{2+3+4}{10+15+20} = \frac{9}{45} = \frac{1}{5}$)

To show that **8** is true, see Appendix, p. 412

Problems Involving Proportion

Example 1 Find x from the proportion x is to 4 as 3 is to 2.

Solution

$$\frac{x}{4} = \frac{3}{2}$$

$$2x = 12 \qquad \text{(cross multiplication)}$$

$$x = 6$$

(You could also solve the above equation by multiplying both sides of the equation by 4)

Example 2 Find x from the proportion: 6 is to x as 4 is to 12.

Solution

$$\frac{6}{x} = \frac{4}{12}$$

$$4x = 72; \text{ and } x = 18$$

Simple Inverse Proportion (Inverse Variation) 6 1

Discussion: Another type of a relationship called **inverse proportion** (or indirect proportion) also occurs in everyday life. For example, suppose that when you drive your car from City A to City B at a speed of 25 mph, it takes you 4 hours. Now, if you always drive at a speed of 25 mph from City A to City B, and it takes you 4 hours, then only two quantities are involved, namely 25 mph and 4 hours. However, if you want to travel the same distance at a speed of say, 50 mph, then you would like to know the time needed to drive from City A to City B at 50 mph, and in this case, four quantities would be involved in the problem, namely, 25 mph, 4 hours, 50 mph, and the time, $t,$ for 50 mph. Since for the same distance, as the speed is increased, the time taken decreases, or if the speed is decreased, the time taken increases, speed is said to be inversely proportional to the time.

Recall the main difference between direct proportion and inverse proportion: for direct proportion, both quantities increase or decrease at the same time; whereas for inverse proportion, as one quantity increases, the other quantity decreases and vice versa.
For example, as one quantity is doubled, the other quantity is halved; but if one quantity is halved, the other quantity is doubled.

Example 1 The time taken by a number of people to do a piece of work is **inversely proportional** to the number of people (assuming that each person works at the same rate as everyone else). Thus, as the number of people increases, the time taken to do the work decreases but as the number of people decreases, the time taken increases (i.e., more people, less time; and less people, more time)

Example 2 At constant temperature, the volume of a given mass of a gas is inversely proportional to the pressure on the gas. (This relationship is known as Boyle's law.)

Definition

If a is inversely proportional to b. (same as a is proportional to the **reciprocal** of b)

      Then a_1 corresponds to b_1, and a_2 corresponds to b_2. and

$$a_1 : \frac{1}{b_1} = a_2 : \frac{1}{b_2}$$

$$a_1 \div \frac{1}{b_1} = a_2 \div \frac{1}{b_2}$$

$$a_1 b_1 = a_2 b_2 \qquad (\textbf{product = product})$$

    Or if we use the familiar terms a, b, c, d as the terms of the inverse proportion

    Then $a : \frac{1}{b} = c : \frac{1}{d}$

$$a \div \frac{1}{b} = c \div \frac{1}{d}$$

$$ab = cd \qquad (\textbf{product = product}) \text{ <------ inverse proportion}$$

We will use the product = product form since this form is easy to remember

Compare:

Direct proportion $\frac{a}{b} = \frac{c}{d}$ <-------- (**quotient = quotient**)

Indirect proportion $ab = cd$ <------(**product = product**)

Determining the type of Proportion (Variation) from the Wording   62

Explicit Specification of direct proportion (variation)

There are a number of ways of **explicitly** specifying direct proportion (variation)
If y is a function of x, then the following are equivalent to one another:

(a) y is directly proportional to x, or simply
(b) y is proportional to x. (The word "directly" is omitted.)
(c) y varies directly as x, or simply,
(d) y varies as x, (The word "directly" is omitted.)

The methods we will cover for solving direct proportion problems can be used to solve any of the problems worded as in (a), (b), (c) and (d) above.

Explicit Specification of inverse proportion (variation)

If y is a function of x, then the following are equivalent to one another:

(a) y is inversely or indirectly proportional to x.
(b) y varies inversely as x,

Any of the methods (we will cover) for solving indirect proportion problems can be used to solve any of the problems stated as in (a) and (b) above.

Implicit Specification of proportion (variation)

In most problems, the terms " directly proportional to" , "inversely proportional to",
" varies directly as" or "varies inversely as" and similar terms are not used in wording the
 problem; but rather, we have to deduce from the problem whether it is a direct or an indirect
 proportion from experience. For example, if 15 oranges cost \$3, what is the cost of 10 oranges?.
 From life experience, we know that at the same price per an orange, as the **number** of oranges
increases, the **cost** of oranges **increases** accordingly. Therefore, the relationship in this problem
 is that of direct **proportion** (variation).

Lesson 17 Exercises A

A. Solve for x: **1.** $\frac{x}{8} = \frac{3}{2}$;   **2.** $\frac{5}{8} = \frac{20}{x}$;   **3.** $\frac{9}{x} = \frac{3}{5}$;   **4.** $\frac{6}{8} = \frac{x}{2}$;   **5.** $\frac{x}{9} = \frac{3}{2}$;

6. $\frac{10}{x} = \frac{3}{5}$;   **7.** $\frac{x}{8} = \frac{4\frac{1}{2}}{2}$;   **8.** $\frac{x}{8\frac{2}{3}} = \frac{3\frac{2}{3}}{2\frac{3}{4}}$;   **9.** $\frac{x}{1.8} = \frac{3}{2.1}$;   **10.** $\frac{x}{8} = \frac{12\%}{96}$

Answers:**1.** $x = 12$; **2.** $x = 32$; **3.** $x = 15$; **4.** $x = \frac{3}{2}$; **5.** $13\frac{1}{2}$; **6.** $16\frac{2}{3}$; **7.** $x = 18$; **8.** $11\frac{5}{9}$;

9. $2\frac{4}{7}$; **10.** 1% or $\frac{1}{100}$

Lesson 17 Exercises B

Determine which of the following are direct proportion or inverse proportion:

1. If 10 gallons of gasoline cost 15 dollars, find the cost of 24 gallons of gasoline.

2. If on 2 gallons of gasoline, a car can travel 24 miles, how many miles can a car travel on 20 gallons of gasoline?

3. If 8 oranges sell for 2 dollars, find the selling price for 24 oranges.

4. If 2 people can paint a house in 6 days, how many people would paint the house in 3 days?

5. At a speed of 60 miles per hour, a car takes 2 hours to travel from City A to City B. At a speed of 40 miles per hour, how long will it take the car to ravel from City A to City B?

6. If 6 mail men can sort a given quantity of mail in 4 hours, how many mail men would be needed to sort the same quantity of mail in 2 hours?

Answers , 1, 2, 3 are direct proportion; 4, 5 and 6 are inverse proportion

Lesson 18

Methods for Solving Direct Proportion Problems
(Direct Variation Problems)

There are a number of methods or approaches for solving **direct proportion** problems. In simple proportion problems, four quantities are involved, and we are (usually) given three quantities and we are required to find the fourth quantity.

We cover **five** methods with examples, and then discuss the relative merits of the methods.

Method 1: Quotient = Quotient or Ratio = Ratio method (Algebraic)

In this method, we word or reword the problem as discussed below.

Consider a direct proportion problem involving four quantities such that a_1 corresponds to b_1, and a_2 corresponds to b_2. We word the proportion as follows:

The proportion a_1 is to b_1 as a_2 is to b_2 translates into the equation

$$\frac{a_1}{b_1} = \frac{a_2}{b_2} \quad \textbf{(quotient = quotient or } \text{ratio} = \text{ratio)} \textbf{ or}$$

Using the familiar notation "a is to b as c is to d", we obtain

$$\frac{a}{b} = \frac{c}{d} \quad \textbf{(quotient = quotient)}$$

Example 1

If 2 dollars can buy 8 apples, how many dollars are needed to buy 24 apples?

Method 1: Quotient = Quotient Method (Algebraic)

We are required to find the number of dollars needed to buy 24 apples.

Step 1: Let the number of dollars needed to buy 24 apples $= x$.

    (we represent the unknown by a variable)

    2 dollars correspond to 8 apples

    x dollars correspond to 24 apples.

    (Note the order in which the units in the second correspondence is mentioned. That is, dollars before apples, and not apples before dollars)

Step 2: 2 dollars are to 8 apples as x dollars are to 24 apples

Step 3: Translating into an equation, $\dfrac{2 \text{ dollars}}{8 \text{ apples}} = \dfrac{x \text{ dollars}}{24 \text{ apples}}$

Step 4: Cross-multiplying, $8x = 2(24)$ ($8x$(dollars)(appless) = 2(24)(dollars)(apples))

Step 5 Solve for x: $x = \dfrac{2(24)}{8}$ (Note above that the units cancel out)

         $= 6$

Conclusion: 6 dollars are needed to buy 24 apples.

Note in the above example that the order in which we mention the units, dollars and apples is important. Note also, however, that in the above example, we could word the proportion as follows:

The proportion, 8 apples are to 2 dollars as 24 apples are to x dollars" translates to $\dfrac{8}{2} = \dfrac{24}{x}$

$$8x = 2(24)$$
$$x = \frac{2(24)}{8}$$
$$= 6 \quad \text{(Again, we obtain the same solution as before.)}$$

Method 2 Units Label Method or Dimensions Method   (Arithmetic method)
This method is highly recommended in applications such as converting from one unit to another unit in measurements as well as calculations in chemistry. In Chemistry, some authors call this method " The factor label method". In this method, we use the units of the given quantities to guide us in obtaining an expression which on simplification yields the desired result.

We repeat the question: If 2 dollars can buy 8 apples, how many dollars are needed to buy 24 apples?

Solution: We can consider this problem as converting 24 apples to dollars. The justification of this method is in parentheses.

Step 1: $\dfrac{24 \text{ apples}}{1} \times \dfrac{?}{?}$

Step 2: Multiply $\dfrac{24 \text{ apples}}{1} \times \dfrac{?}{?}$ by a fraction formed by using the quantities 2 dollars and

8 apples as the terms of the fraction such that the denominator has the same units as the 24 apples (the quantity in the numerator), and under such conditions, we can divide out (cancel) the common units in the numerator and the denominator, leaving us the units, dollars.

Then, $\dfrac{24 \text{ apples}}{1} \times \dfrac{2 \text{ dollars}}{8 \text{ apples}}$

Step 3: $\dfrac{24 \ \cancel{\text{apples}}}{1} \times \dfrac{2 \text{ dollars}}{8 \ \cancel{\text{apples}}}$ (= Number of apples ✕ cost per each apple)

   = 6 dollars

Conclusion: 6 dollars are needed to buy 24 apples.

Let us do the above problem backwards, with the justification of this method in parentheses
If 2 dollars can buy 8 apples, how many apples can one buy for 6 dollars?
Solution Here, we are converting 6 dollars to apples.

   $\dfrac{6 \text{ dollars}}{1} \times \dfrac{?}{?}$

= $\dfrac{6 \text{ dollars}}{1} \times \dfrac{8 \text{ apples}}{2 \text{ dollars}}$ $\left(= \text{total cost} \div \text{cost per 1 apple} = \text{total cost} \ ✕ \ \dfrac{1}{\text{cost per 1 apple}}\right)$

= 24 apples, which is what was given in the original problem.

Method 3 Unitary Method (Arithmetic)
We repeat the question: If 2 dollars can buy 8 apples, how many dollars are needed to buy 24 apples?

Solution: Cost of 8 apples = 2 dollars

   Cost of 1 apple = $\dfrac{2 \text{ dollars}}{8}$ (Divide to obtain the cost for 1 apple))

   Cost of 24 apples = $\dfrac{2}{8} \times \dfrac{24}{1}$ dollars (Multiply to obtain the cost for other than 1)

   = 6 dollars

Conclusion: 6 dollars are needed to buy 24 apples.

Method 4 "More or Less Method" (Arithmetic method. This method the shorter form of Method 3.
We repeat the question: If 2 dollars can buy 8 apples, how many dollars are needed to buy
24 apple?

If cost of 8 apples = 2 dollars

Then cost of 24 apples = $\dfrac{2 \text{ dollars}}{1} \times \dfrac{24 \text{ apples}}{8 \text{ apples}}$

= 6 dollars

Conclusion: 6 dollars are needed to buy 24 apples.

In method 4, we used a very useful principle which states that " If more, the smaller divides, and if less, the larger
divides". That is, since we expect 24 apples to cost more than 2 dollars, in forming the fraction involving 8
apples and 24 apples, the smaller of 8 apples and 24 apples is the divisor (smaller divides), and hence we used
the fraction,

$\dfrac{24 \text{ apples}}{8 \text{ apples}}$ <----- " smaller divides" (We multiplied by an i**mproper fraction**)

However, if we had expected the number (answer) to be less than 2 dollars, we would have used the fraction
$\dfrac{8 \text{ apples}}{24 \text{ apples}}$ <----- "larger divides" (i.e., we would have multiplied by a **proper fraction**)

The More or Less Method above is very powerful once you have mastered it, because it can be applied to inverse
proportion problems as well as to problems involving both inverse and direct proportion and any number
of quantities (variables), especially in science.

Method 5 Variation Method (Algebraic)

We reword the problem: The number of apples a person buys varies directly as the number of dollars
paid. If when the number of apples is 8, the number of dollars paid is 2, find the number of
dollars paid when the number of apples is 24.

Step 1: Let the number of apples = A
Let the number of dollars paid = D
Then $A = kD$, where k is called the proportionality constant.
Step 2: Substituting $A = 8, D = 2$ *in* $A = kD$,
$8 = 2k$
$4 = k$
Substituting $k = 4$ *in* $A = $ kD, we obtain the formula,
$A = 4D$.
Step 3: When $A = 24$
$24 = 4D$ (substituting 24 for A in the formula, $A = 4D$)
$6 = D$ (solving for D)
$D = 6$
Therefore, 6 dollars would be paid for 24 apples.

Example 2 On 3 gallons of gasoline, one can drive 60 miles. How far can one drive on 8
gallons of gasoline?
Solution
Method 1 Quotient = Quotient Method (Algebraic)
Let the number of miles for 8 gallons of gasoline = x
Proportion statement: 3 gallons are to 60 miles as 8 gallons are to x miles.

Translating, $\dfrac{3 \text{ gallons}}{60 \text{ miles}} = \dfrac{8 \text{ gallons}}{x \text{ miles}}$

$\dfrac{3}{60} = \dfrac{8}{x}$

$3x = 60(8)$

$$x = \frac{60(8)}{3}$$

$$x = 160$$

Therefore, on 8 gallons of gasoline, one can drive 160 miles.

Note above that the ratio, gallons to miles = the ratio, gallons to miles; or
  the ratio miles to gallons = the ratio miles to gallons
Incorrect: the ratio miles to gallons = the ratio gallons to miles. (violation of the order of the units)

All other proportion problems involving four quantities can be reworded and translated into equations by imitating the examples covered.. Only, the units involved might be different. We may have apples and pounds as the units involved, or hours and dollars, or years and people.

Method 2 Units Label Method. **We repeat the problem:** On 3 gallons of gasoline, one can drive 60 miles. How far can one drive on 8 gallons of gasoline?

Solution: We can consider this problem as converting 8 gallons to miles.

Step 1: $\dfrac{8 \text{ gallons}}{1} \times \dfrac{?}{?}$

Step 2: Multiply $\dfrac{8 \text{ gallons}}{1} \times \dfrac{?}{?}$ by a fraction formed by using the quantities 3 gallons and 60 miles

$\dfrac{8 \text{ gallons}}{1} \times \dfrac{60 \text{ miles}}{3 \text{ gallons}}$

Step 3: $\dfrac{8 \text{ \sout{gallons}}}{1} \times \dfrac{60 \text{ miles}}{3 \text{ \sout{gallons}}}$

$= 160 \text{ miles}$

Therefore, on 8 gallons of gasoline, one can drive 160 miles.

Method 3 Unitary Method (Arithmetic)

We repeat the problem: On 3 gallons of gasoline, one can drive 60 miles.
  How far can one drive on 8 gallons of gasoline?

Solution

Number of miles driven on 3 gallons of gasoline = 60 miles

Number of miles driven on 1 gallon of gasoline = $\frac{60}{3}$ miles

(Divide to obtain the miles for 1 gallon)

Number of miles driven on 8 gallon of gasoline = $\frac{60}{3} \times 8$ miles (Multiply for other than 1 gallon)

$= 160 \text{ miles}$

Therefore, on 8 gallons of gasoline, one can drive 160 miles.

Method 4 More or Less Method **We repeat the problem:**

On 3 gallons of gasoline, one can drive 60 miles. How far can one drive on 8 gallons of gasoline?

Solution   Number of miles driven on 3 gallons gasoline = 60 miles

Number of miles driven on 8 gallons gasoline = $\frac{60}{1} \times \frac{8}{3}$ miles

$= 160 \text{ miles}$

Therefore, on 8 gallons of gasoline, one can drive 160 miles.

Method 5 Variation Method (Algebraic) We reword the problem:

> The number of miles one drives is directly proportional to the number
> of gallons of gasoline used. If when the number of gallons of gasoline
> is 3, the number of miles is 60, find the number of miles when the number of gallons
> of gasoline is 8.

Solution

Step 1: Let the number of gallons of gasoline = n

Let the number of miles driven = d

Then since d is directly proportional to n,

$d = kn$, where k is the proportionality constant

Step 2: Substitute for $d = 60$, $n = 3$, in $d = kn$ and solve for k.

Then $60 = k(3)$ and from which $k = 20$

Step 3: Substitute for $k = 20$ in $d = kn$ to obtain a formula relating d and n.

Then $d = 20n$

Step 4: Substitute for $n = 8$ in $d = 20n$ and determine d

$d = (20)(8)$

$\quad = 160$

Therefore, on 8 gallons of gasoline, one can drive 160 miles.

Example 3   The ratio of an astronaut's weight on the earth to weight on the moon is 6:1. If
an astronaut weighs 180 lb (on earth), what is the astronaut's weight on the moon?

Solution

Method 1 Quotient = Quotient Method (Algebraic)

Let the weight on the moon be x lb, (i.e., x lb corresponds to 180 lb on the earth).

Step 1:   The ratio of the weight on earth to weight on the moon is 6:1 or $\frac{6}{1}$.

Step 2:   Also, the ratio of the weight on earth to the weight on the moon is 180 lb to x lb

or $\frac{180}{x}$.

Step 3 :   Equate the two ratios and solve for x.

$$\frac{6}{1} = \frac{180}{x}$$

$$6x = 180(1)$$

$$x = \frac{180(1)}{6}$$

$$x = 30$$

Therefore, the astronaut's weight on the moon is 30 lb.

Method 2 Reword the original problem as a proportion:

6 is to 1 as 180 is to x.

Solution

Translating, $\frac{6}{1} = \frac{180}{x}$

Solving, $x = 30$, and we obtain the same solution as by Methods 1 and 2.

Method 3 Unitary Method (Arithmetic)

Whenever the astronaut weighs 6 lb on the earth, the astronaut weighs 1 lb on the moon.

Whenever the astronaut weighs 1 lb on the earth, the astronaut weighs $\frac{1}{6}$ lb on the moon; and

Whenever the astronaut weighs 180 lb on the earth, the astronaut weighs $\frac{1}{6}$ × 180 lb = 30 lb on the moon.

Method 4: More or Less Method We reword the problem

The weight of an astronaut on earth is directly proportional to the weight on the moon. When the weight on earth is 6 lb, the weight on the moon is 1 lb, find the weight on the moon when the weight on earth is 180 lb..

Solution Weight on the moon when the weight on earth is 6 lb = 1 lb

Weight on moon when weight on earth is 180 lb = 1 ℓb × $\dfrac{180\ \ell b}{60\ \ell b}$ = 30 lb

Therefore, the astronaut 's weight on the moon is 30 lb.

Method 5   Variation Method (Algebraic)

We reword the problem   The weight of an astronaut on earth is directly proportional to the weight on the moon. When the weight on earth is 6 lb, the weight on the moon is 1 lb, find the weight on the moon when the weight on earth is 180 lb..

Solution

Step 1: Let the weight on earth = h lb

Let the weight on moon = n lb

Then since h is directly proportional to n,

$h = kn$

Step 2: Determine the value of k by substituting $h = 6, n = 1$ in $h = kn$

Then $6 = k(1)$ and from which $k = 6$

Step 3: Substitute $k = 6$ in $h = kn$ to obtain a formula relating h and n.

Then we obtain $h = 6n$

Step 4: Substitute $h = 180$ in the formula, $h = 6n$ and solve for n.

Then $180 = 6n$

$30 = n$

Therefore, the astronaut 's weight on the moon is 30 lb.

More Examples on the Variation Method

(Note that these examples can also be solved using Methods 1, 2, 3 and 4)

Example A: y varies directly as x. When $y = 48, x = 3$. Find x when $y = 96$

Solution

Step 1: $y = kx$ (1)

Step 2: Substitute $y = 40, x = 2$ in equation (1) and solve for k.

$48 = k(3)$

$48 = 3k$

$16 = k$

Substitute $k = 16$ in equation (1) above to obtain

the formula $y = 16x$ (2)

Step 3: Now, to find x when $y = 96$, replace y by 96 in (2)

Then $96 = 16x$

$6 = x$

Therefore when $y = 96, x = 6$.

Example B The distance y an object falls from rest is directly proportional to the square of the time x. When $y = 128, x = 4$. *(a)* Find x when $y = 200'$ *(b)* Find y when $x = 6$.

Solution

Step 1: $y = kx^2$ ....(1)

Step 2: Substitute $y = 128, x = 4$ in equation (1) and solve for k.

Then $128 = k(4)^2$

$128 = 16k$

$k = 8$

Substitute $k = 8$ in equation (1) above to obtain

the formula $y = 8x^2$ .... (2)

Step 3: (a) To find x when $y = 200$. Replace y by 200 in equation (2) and solve for x.

$200 = 8x^2$

$\dfrac{200}{8} = x^2$

$25 = x^2$

$5 = x$

$x = 5$ (Since the time taken must be positive in this problem, we reject -5 as a solution)

(b) To find y when $x = 6$, replace x by 6 in equation (2) and evaluate.

$y = 8(6)^2$

$y = 8(36)$

$y = 288$.

Therefore when $x = 6, y = 288$.

EXTRA

Example 4 Assume volume V is directly proportion to temperature T.

Then V_1 corresponds to T_1 as V_2 corresponds to T_2

Since this is direct proportion, we apply **quotient = quotient**

Then $\dfrac{V_1}{T_1} = \dfrac{V_2}{T_2}$ (quotient = quotient)

The above equation is a familiar formula for gases in physics.

Relative Merits of the Methods for Solving Proportion Problems 71

Method 1: Quotient = quotient Method

The quotient = quotient method such as $\frac{a}{b} = \frac{c}{d}$ is very popular and is efficient for a one-step problem involving only four quantities, but is not efficient in a problem such as in Method 2 below, since the method has to be repeated in steps with the introduction of a new variable in each step.

Method 2: Units Label Method (Arithmetic)

The Units Label Method, an arithmetic method, can readily do what Method 1 can do, if the quantities involved have units, and is more efficient than Method 1 in doing repeated calculations. This method is particularly very efficient in calculations involving measurements and in chemistry as in the following example.

Example How many yards are there in 96 inches?

Solution We reword this problem as: Convert 96 in. to yd.

Plan: This plan is based on what is available in the conversion chart or table.
(Using a conversion chart is analogous to using a map to go from one location to another)
The sequence is inches--> feet---> yards.
We use these relationships: 12 in. = 1 ft;   3 ft = 1 yd (We could use 36 in = 1 yd)
 (Note: inch = in.)

Step 1: $\frac{96 \text{ in.}}{1} \times \frac{?}{?}$

Step 2: Multiply the $\frac{96 \text{ in.}}{1}$ by a fraction formed by using the quantities 12 in. and 1 ft as the terms of the fraction such that the denominator has the same units as the 96 in.; followed by similar multiplication by a fraction using 3 ft and 1 yd as the terms of the fraction. Then, we obtain

$$\frac{96 \text{ in}}{1} \times \frac{1 \text{ ft}}{12 \text{ in}} \times \frac{1 \text{ yd}}{3 \text{ ft}} \qquad \left(\frac{96 \text{ in}}{1} \times \frac{1 \text{ ft}}{12 \text{ in}} = \frac{96 \text{ in}}{1} \times \frac{1 \text{ ft}}{12 \text{ in}} \times \frac{1 \text{ yd}}{3 \text{ ft}} = \frac{96 \times 1 \times 1 \text{ yd}}{1 \times 12 \times 3} \right)$$

$$= 2\tfrac{2}{3} \text{ yd}$$

$$\therefore \ 96 \text{ in.} = 2\tfrac{2}{3} \text{ yd}$$

Justification of the Units Label Method in the above problem:
 From 12 in. = 1 ft, If we divide both sides of this equality by 12 in., we obtain

$$\frac{12 \text{ in.}}{12 \text{ in.}} = \frac{1 \text{ ft}}{12 \text{ in.}}$$

$$1 = \frac{1 \text{ ft.}}{12 \text{ in.}}$$

Similarly, from 3 ft = 1 yd, If we divide both sides by 3 ft., we obtain

$$\frac{3 \text{ ft}}{3 \text{ ft}} = \frac{1 \text{ yd}}{3 \text{ ft}}$$

$$1 = \frac{1 \text{ yd}}{3 \text{ ft}}$$

In step 2, multiplying by $\frac{1 \text{ ft.}}{12 \text{ in.}}$ or by $\frac{1 \text{ yd}}{3 \text{ ft}}$ is equivalent to multiplying by 1, and therefore, the given value remains equivalent.

Note: In the above example, if we were to use the quotient = quotient method, we would have to apply this method twice, with the introduction of a variable in each step. Note above also that it is easy to generalize the justification of the "Units Label Method" when we have equality such as 12 in. = 1 ft, than when we have equivalence such as 8 apples are equivalent to 2 dollars.

Method 3: Unitary Method (Arithmetic)
This is an arithmetic method and does not require any knowledge of algebra and can be taught
 as the first arithmetic method for solving proportion problems.

Method 4: More or Less Method (Arithmetic)
This method is a shorter form of Method 3, as one step in Method 3 is skipped. When mastered
well, the "More or Less Method" can be extended to solve inverse proportion as well as compound
proportion problems, and very efficient in stoichiometric calculations in chemistry as well as
conversions in measurements.

Method 4: Variation Method (Algebraic)
This has applications at all levels of mathematics. Some textbooks ignore this method at the elementary
level and cover it at the intermediate level.

EXTRA: Two Cases of the Statement of Ratio and the Corresponding Proportion Problems

In solving problems involving ratios and proportion, we must distinguish between the cases in which the terms of a given ratio refer to **disjoint sets** and the case in which the ratios refer to sets which are **not** disjoint.

Case 1
The terms of the ratios refer to disjoint sets (example: the set of females and the set of males in a school)
Example: The ratio of the number of females to the number of males is 3 to 2.
Proportion: The ratio 3 **females** to 2 **males** = the ratio 6 **females** to 4 **males**
       or ratio 2 **males** to 3 **females** = ratio 4 **males** to 6 **females**
(Two sets are disjoint if the sets do not have any members common to both sets)
Problems such as " If 8 apples cost 2 dollars, what is the cost of 24 apples" are of case 1

Case 2
The terms do **not r**efer to disjoint sets (example: the set of females and the set of the total number of students at a school)

Example
The ratio of the number of **females** to the **total number** of students is 3 to 5.
Proportion: The ratio 3 **females** to 5 **students** = the ratio 6 **females** to 10 **students.**
Case 2 is sometimes worded as **a number** out of a **total number**
For example, 3 out of 5 students are females or out of every 5 students, 3 are females,
which is the same as $\frac{3}{5}$ of the students are females.
Problems such as " if 3 out of 5 students in a school are females, and there are 600 females in this school, how many students are in this school?" can also be set up using
$$\frac{\text{what number out of}}{\text{what number}} = \frac{\text{what number out of}}{\text{what number}} \text{ as well as other methods.}$$

Example for Case 2
Example 1   3 out of 5 students at a certain college are females.

        (a) If 600 students at this college are females,
            how many students are there at this college?

        (b) If one were to collect a sample of 200 students at this
            college, how many of these students would be females?

Solution (a):
Method 1 Using ratio = ratio method
           Let the number of students at this college = x
Then the proportion 3 females are to 5 students as 600 females are to x students
translates to: $\dfrac{3 \text{ females}}{5 \text{ students}} = \dfrac{600 \text{ females}}{x \text{ students}}$

$$\frac{3}{5} = \frac{600}{x}$$
$$3x = 5(600)$$
$$x = \frac{5(600)}{3} ; \ x = 1000$$

Therefore, there are 1000 students at this college.

Method 2: Using $\dfrac{\text{what number out of}}{\text{what number}} = \dfrac{\text{what number out of}}{\text{what number}}$

Let the number of students at this college = x

$$\dfrac{3 \text{ out of}}{5} = \dfrac{600 \text{ out of}}{x}$$

$$\dfrac{3}{5} = \dfrac{600}{x}$$

$$3x = 5(600)$$

$$x = \dfrac{5(600)}{3}$$

$$x = 1000$$

Therefore, there are 1000 students at this college

Method 3   Since 3 out of 5 means $\dfrac{3}{5}$, we can reword this part of the question as: if $\dfrac{3}{5}$ of a number is 600, what is the number?

Let the number be x

Then $\dfrac{3x}{5} = 600$

$$3x = 3000$$

$$x = 1000$$

There are 1000 students at this college.

Method 4 Divide 600 by $\dfrac{3}{5}$ (If $\dfrac{3}{5}$ of a number is 600, what is the number?)

$$600 \div \dfrac{3}{5} = \dfrac{600}{1} \times \dfrac{5}{3}$$

$$= 1000$$

There are 1000 students at this college.

Method 5: The Units Label Method

Note: 3 out 5 students are females means that for every 5 students, there are 3 females.

Step 1: $\dfrac{600 \text{ females}}{1} \times \dfrac{?}{?}$

Step 2: $= \dfrac{600 \text{ females}}{1} \times \dfrac{5 \text{ students}}{3 \text{ females}}$

$= 1000$ students

Part (b)

Method 1:   We can reword this part of the question as: Find $\dfrac{3}{5}$ of 200.

$$\dfrac{3}{5} \text{ of } 200 = \dfrac{3}{5} \times \dfrac{200}{1} = 120$$

That is, of a sample of 200 students, 120 students would be females.

Method 2:: $\dfrac{\text{what number out of}}{\text{what number}} = \dfrac{\text{what number out of}}{\text{what number}}$

Let the number of females $= x$

$$\frac{3 \text{ out of}}{5} = \frac{x \text{ out of}}{200}$$

$$\frac{3}{5} = \frac{x}{200}$$

$$5x = 3(200)$$

$$x = \frac{3(200)}{5}$$

$$x = 120$$

Therefore, there are 120 females.

Method 3: The Units Label Method

Step 1: $\dfrac{200 \text{ students}}{1} \times \dfrac{?}{?}$

Step 2: $\dfrac{200 \text{ students}}{1} \times \dfrac{3 \text{ females}}{5 \text{ students}}$

$\qquad = 120 \text{ females}$

Therefore, there are 120 females.

Method 4

Let the number of females. $= x$

Then the proportion 3 females are to 5 students as x females are to 200 students translates to:

$$\frac{3 \text{ females}}{5 \text{ students}} = \frac{x \text{ females}}{200 \text{ students}}$$

$$\frac{3}{5} = \frac{x}{200}$$

$$5x = 3(200)$$

$$x = \frac{3(200)}{5} = 120$$

Therefore, there are 120 females..

Extra Method

We repeat the question in **Example 1**   3 out of 5 students at a certain college are females.

(a) If 600 students at this college are females,
     how many students are there at this college?

(b) If one were to collect a sample of 200 students at this
     college, how many of these students would be females?

This method follows the algebraic method for dividing a quantity into parts, given the ratio.
Since the ratio of females to the total number of students is 3:5,
let the number of females $= 3x$, and
let the total number of students $= 5x$.
(a) Then $3x = 600$ (there are 600 females); and solving, $x = 200$
Therefore, total number of students $= 5x = 5(200) = $**1000**
Therefore, there are 1000 students at this college.

(b) $5x = 200$ (there are 200 students) , and
$\qquad x = 40$
Therefore, the number of females $= 3x = 3(40) = 120$.

Lesson 18 Exercises A

Problems on direct proportion in which the terms refer to distinct sets

1. If 1 gallon of gasoline costs 2 dollars, how many gallons of gasoline can one buy for 15 dollars?　　　　　　　　　　　　Answer: $7\frac{1}{2}$ gallons

2. If 10 gallons of gasoline cost 15 dollars, find the cost of 24 gallons of gasoline.
   　　　　　　　　　　　　Answer: 36 dollars

3. If 10 gallons of gasoline cost 15 dollars, how many gallons of gasoline can one buy for 45 dollars?　　　　　　　Answer: 30 gallons

4. If x gallons of gasoline cost y dollars, what is the cost of p gallons of gasoline
   (Express your answer in terms of x y and p)　　　　Answer: $\frac{py}{x}$ dollars

5. If on a gallon of gasoline, a car can travel 24 miles, how many miles can the car travel on 10 gallons of gasoline?　　　　　　Answer: 240 miles

6. If on 2 gallons of gasoline, a car can travel 24 miles, how many miles can a car travel on 20 gallons of gasoline?　　　　　　Answer: 240 miles

7. If on 3 gallons of gasoline, a bus can travel 36 miles, how many gallons of gasoline are needed to travel 60 miles?　　　　　Answer: 5 gallons

8. If on p gallons of gasoline, a bus can travel r miles, how many gallons of gasoline are needed to travel m miles? (Express your answer in terms of the variables.)　　Answer: $\frac{mp}{r}$ gallons

9. If 4 oranges cost 1 dollar, how many oranges can one buy for 20 dollars?
   　　　　　　　　　　　　Answer: 80 oranges

10. If 8 oranges sell for 2 dollars, find the selling price for 24 oranges.　　Answer: 6 dollars

11. If 12 oranges cost 3 dollars, what is the cost of 60 oranges?　　Answer: 15 dollars

12. If r oranges cost p dollars, what is the cost of m oranges?　　Answer: $\frac{mp}{r}$ dollars
    (Express your answer in terms of the variables.)

13. If a household uses one pound of sugar in 3 weeks, how many pounds of sugar would be used in 6 weeks?　　　　　　Answer: 2 pounds

14. If a household uses 2 pounds of sugar in 3 weeks, how many pounds of sugar would be used in 51 weeks?　　　　　　Answer: 34 pounds

15. If a household uses 2 pounds of sugar in 3 weeks, 12 pounds of sugar would be used in how many weeks?　　　　　　Answer: 18 weeks

16. If a household uses x pounds of sugar in r weeks, how many pounds of sugar would be used in y weeks? (Express your answer in terms of the variables.)　　Answer $\frac{xy}{r}$ pounds

17. If one teacher is needed for a class of 22 students, how many teachers would be needed to teach 88 students, assuming that each class has 22 students?   Answer: 4 teachers

18. If 6 teachers are needed to teach 102 students, how many students would be taught by 3 teachers?   Answer: 51 students

19. If t teachers are needed to teach s students, how many students would be taught by b teachers? (Express your answer in terms of the variables.)   Answer: $\frac{bs}{t}$ students

20. If 4 grams of hydrogen are required to produce 36 grams of water, how many grams of hydrogen would be required to produce 108 grams of water?   Answer:: 12 grams

21, If from 4 grams of hydrogen, 36 grams of water are produced, how many grams of hydrogen are needed to produce 216 grams of water?   Answer: 24 grams

22. If from t grams of hydrogen, s grams of water are produced, how many grams of hydrogen are needed to produce b grams of water (Express your answer in terms of the variables   Answer: $\frac{bt}{s}$ grams

23. If p gallons of water are needed to produce x pounds of a concrete mixture, how many gallons of water are needed to produce y pounds of this mixture?   Answer: $\frac{py}{x}$ gallons

24. 1f each room in a building has 3 windows, how many windows are there in 6 rooms,?   Answer: 18 weeks

25. If 3 rooms in a building have 12 windows, how many windows are in 36 rooms, assuming that each room has the same number of windows?
Answer: 144 windows

26. If p rooms in a building have q windows, how many windows are in r rooms, assuming that each room has the same number of windows? (Express answer in terms of the variables.)
Answer: $\frac{qr}{p}$

27. A buyer purchased 20 items for 60 dollars. How many items can be purchased for 30 dollars, assuming that the cost is the same for each of the items.   Answer: 10 items

28. A buyer purchased r items for s dollars. How many items can be purchased for t dollars, assuming that the cost is the same for each of the items. (Answer in terms of the variables.)
Answer: $\frac{rt}{s}$

29. In 2 hours, a car can travel 120 miles. Find the distance the car can travel in 3 hours, assuming the same speed   Answer:: 180 miles

30. In x hours, a car can travel y miles. Find the distance the car can travel in k hours, assuming the same speed.   Answer: $\frac{ky}{x}$

31. In 2 hours, a car can travel 120 miles, At the same speed, how long will it take the car to travel 180 miles?.   Answer: 3 hours

32. In x hours, a car can travel y miles. At the same speed, how long will take the car to travel t miles? (Express your answer in terms of the variables)   Answer: $\frac{xt}{y}$

33. If 40 liters of air are needed to ventilate a patient for 2 minutes, how many liters of air are needed to ventilate this patient for 24 minutes? Answer: 480 liters

34. If a liters of air are needed to ventilate a patient for b minutes, how many liters of air are needed to ventilate this patient for c minutes? Answer $\frac{ac}{b}$

35. If $4\frac{1}{2}$ gallons of gasoline cost 6 dollars, how many gallons of gasoline can one buy for 16 dollars?   Answer: 12 gallons

36. If there are 12 inches in 1 foot, how many inches are there in 4 feet?   Answer: 48 inches

37. If there are 12 inches in I foot, how many inches are there in x feet?
(answer in terms of x) Answer: $12x$ inches

38. If a person's heart beats 4 times every 3 seconds, how many times will it beat in 21 seconds.
Answer: 28 times

39. To prepare an optimum mixture of gasohol which consists of alcohol and gasoline, Alexander used 20 liters of alcohol and 180 liters of gasoline. How many liters of alcohol are needed if 270 liters of gasoline are used and the composition ratio of the mixture remains unchanged?   Answer: 30 liters

40. Charles' Law states that at constant pressure, the volume of a gas is directly proportional the temperature of the gas. If when the volume of a given mass of a gas is 4.2 L, the temperature is $309°K$, find the volume when the temperature is $297°K$ Answer : 4.0 L

41. Gay-Lussac's Law states that at constant volume, the pressure on a gas is directly proportional to the temperature of the gas. If when the pressure on a given mass of a gas is 720 mm Hg, the temperature is $303°K$, find the pressure when the temperature is $309°K$?   Answer: 734 mm Hg

4 2 The ratio of a man's weight on the earth to his weight on the moon is 6:1. If the man weighs (on earth) 240 lb, what is his weight on the moon? Answer: 40 l

Lesson 18 Exercises B
Problems on direct proportion in which some terms refer to non distinct sets

1. If 2 out of 5 students at a college are males and there are 2,800 students at this college, how many of the students are males?   Answers: 1,120

2. If 2 out of 5 students at a college are males and there are 2,800 students at this college, how many of the students are females?   Answers: 1,680

3. If 2 out of 5 students at a college are males and there are 1,120 males at this college, how many students are at this college?   Answers: 2,800

4.. If 40% of students at a college are males and there are 1,120 males at this college, how   many students are at this college?   Answers: 2,800

5. If 40% of the students at a college are males and there are 2,800 students at this college, how many of the students are males?   Answers: 1,120

6. Is $\frac{2}{5}$ of a number is 16, what is the number?   Answer: 40

7. James and Mary are to divide some amount of money and James is to receive $\frac{5}{12}$ of this amount. If James receives 30 dollars, what is the amount of money to be divided?

Answer: 72 dollars

8. If 2 out 3 dollars in a bank account are to be given to Diana, If Diana was given $30,000. how much money is in this bank account?   Answer: $45,000

9. If 1 out of 3 liters of an alcohol-water solution is pure alcohol, and this solution contains 12 liters of pure alcohol, what is the total volume of this alcohol-water solution?

Answer 36 liters

10. If q out of p fruits are oranges, and there are r oranges, how many fruits are there. Obtain an expression for the number of fruits in terms of p, q, and r. Answer: $\frac{pr}{q}$

11. In a math class, 2 out of 4 students passed the final exam. If 32 students passed the exam, how many students are in this class?   Answer: 64

Lesson 18 Exercises C

1. y varies as x. When $y = 12$, $x = 4$. Find y when $x = 10$.

2. The distance S an object falls from rest is directly proportional to the square of the time t. If when $S = 64$, $t = 2$, find S when $t = 4$.

3. W is proportional to V. When $W = 120$, $V = 3$, find W when $V = 8$.

4.. S is directly proportional to t.. When $S = 64$, $t = 2$. Find t when $S = 288$.

5. Express by means of an equation: $(a - b)$ is proportional to $(c - d)$.

6. The distance S an object falls from rest varies directly as the square of the time t. If when $S = 32$, $t = 1$, find S when $t = 5$.

Answers: 1.   30;   **2.** 256; **3.** 320; **4.** 9;   **5.** $a - b = k(c - d)$, where k is a constant; **6.** 800.

Lesson 18 Exercises D

1. At constant pressure, the volume V of a perfect gas varies as the absolute temperature T.

   If when the temperature is 2730 Absolute, the volume of one gram of a gas is 1.7 cm^3, find the volume when the absolute temperature is 4830°.

2. According to Hooke's law, the force required to stretch a spring is directly proportional to the elongation of the spring. If a 20 lb force stretches a spring 6 in., what force will be required to stretch it 8 in.?

3. The cost C of gasoline is proportional to the number of gallons N of gasoline. If 10 gallons of gasoline cost $12, what is the cost of 25 gallons of gasoline? How many gallons of gasoline can $18 purchase?

Answers: 1. 3.0 cm^3; **2.** 27 lb; **3.** $30; 15 gallons

Lesson 19
Methods for Solving Inverse Proportion Problems

It is very interesting but not surprising that the operations in the steps for solving inverse proportion problems are the opposite operations (inverse operations) in the steps for solving direct proportion problems. By operations, we mean multiplication and division are opposite (inverse) operations. Thus if we know the operations in the steps for solving a direct proportion problem, we can deduce the operations in the steps for solving an inverse proportion problem (by using the opposite operations). For every method for solving direct proportion problems, there is a corresponding method for solving inverse proportion problems, with the steps in the inverse method obtained from the direct proportion method by using opposite operations.

Method 1: **Product = Product Method** (Algebraic)

Assume a is inversely proportional to b.

(same as a is proportional to the **reciprocal** of b)

Then a_1 corresponds to b_1, and a_2 corresponds to b_2. and

$$a_1 : \frac{1}{b_1} \text{ as } a_2 : \frac{1}{b_2}$$

$$a_1 \div \frac{1}{b_1} = a_2 \div \frac{1}{b_2}$$

$$a_1 b_1 = a_2 b_2 \qquad \textbf{(product = product)}$$

Or if we use the familiar terms a, b, c, d as the terms of the inverse proportion

Then $a : \frac{1}{b}$ as $c : \frac{1}{d}$

$a \div \frac{1}{b}$ as $c \div \frac{1}{d}$

$$ab = cd \qquad \textbf{(product = product)} \text{ <------ inverse proportion}$$

We will use the product = product form since this form is easy to remember.

Compare: **Direct proportion** $\qquad \frac{a}{b} = \frac{c}{d}$ <------**(quotient = quotient)**

Inverse proportion $\quad ab = cd$ <------**(product = product)**

Example 1 2 people can paint a house in 6 days. How many people would be required to paint the house in 3 days? Derive a formula relating the number of people and the number of days; and use this formula to find the number of people required to paint the house in 3 days.

Method 1 Product = Product Method

It is known that as the number of people increases, the number of days required decreases, and vice versa.

Let the number of people $= P$,

Let the number of days D.

Then P is indirectly proportional to D,

P_1 corresponds indirectly to D_1 as P_2 corresponds indirectly to D_2.

Since this is indirect proportion, $P_1 D_1 = P_2 D_2$.

From the problem, $P_1 = 2, D_1 = 6, P_2 = ?. D_2 = 3$.

Substituting these values in the equation

$$2(6) = P_2(3)$$

$$\frac{2(6)}{3} = P_2$$

$$4 = P_2$$

Therefore, 4 people are required to paint the house in 3 days.

Method 2: Units Label Method for Inverse Proportion   8 1

By reversing the steps in the corresponding direct proportion method (Method 2), we obtain the approach covered below.

We repeat the question: 2 people can paint a house in 6 days. How many people would be required to paint the house in 3 days?

Step 1: Find the product, 2 people • (6 days)

Step 2: Divide : $\dfrac{2 \text{ people} \bullet (6 \text{ days})}{3 \text{ days}}$

$$\frac{2 \text{ people} \bullet (6\,\text{days})^{2}}{3\,\text{days}}$$

$= 4$

Therefore, 4 people are required to paint the house in 3 days.

Method 3: Unitary Method (Arithmetic)

We repeat the question: 2 people can paint a house in 6 days. How many people would be required to paint the house in 3 days?

Solution

The number of people required for 6 days = 2

The number of people required for 1 day = 2×6 (Multiply. More people are needed)

The number of people required for 3 days = $\dfrac{2 \times 6}{3}$ (We divide. Less people are needed)

$= 4$

Therefore, 4 people are required to paint the house in 3 days.

Method 4: More or Less Method

We repeat the question: 2 people can paint a house in 6 days. How many people would be required to paint the house in 3 days?

Solution

Step 1: The number of people needed to paint the house in 6 days = 2

Step 2: The number of people needed to paint the house in 3 days = $2 \times \dfrac{6}{3}$

$= 4$ (more people needed)

Therefore, 4 people are required to paint the house in 3 days.

(Note that in Step 2, in forming a fraction using 3 and 6, we used an improper faction, $\dfrac{6}{3}$ (fraction greater than or equal to 1), since we expect the number of people to be greater than 2. For the painting to be done faster, more people are needed. If we had expected the number of people to be less than 2, we would have used the proper fraction $\dfrac{3}{6}$). Master this arithmetic method.

Method 5: Variation Method (Algebraic) **We reword the question:**

The number of people needed to paint a house varies inversely as the number of days taken to paint the house. If when the number of people is 2, the number of days is 6, find the number of people when the number of days is 3.

Step 1: Let the number of people = n

Let the number of days = d

Then since n varies inversely as d, (or n is proportional to the **reciprocal** of d

$$n = \frac{k}{d}$$

Step 2: Determine the value of k by substituting $n = 2, d = 6$ in $n = \dfrac{k}{d}$

Then $2 = \dfrac{k}{6}$ and from which $k = 12$.

Step 3: Substitute $k = 12$ in $n = \dfrac{k}{d}$ to obtain a formula relating n and d.

Then we obtain $n = \dfrac{12}{d}$

Step 4: Substitute for $d = 3$ in the formula . $n = \dfrac{12}{d}$ and solve for n

Then $n = \dfrac{12}{3} = 4$

Thus when $d = 3, n = 4$

Therefore, 4 people are required to paint the house in 3 days.

One more example on Method 5

Example 2 The number of people required to sort a quantity of letters varies inversely as the time taken to sort the letters. If 8 people can sort a quantity of letters in 3 hours, how many people would be required to sort the same quantity of letters in 2 hours?

Let the number of people required to sort the letters in x hours be y

Step 1: Then $y = \dfrac{k}{x}$ (1)

Step 2: When $x = 3, y = 8$. Substitute these values in equation (1) and solve for k.

$8 = \dfrac{k}{3}$.

Solving, $k = 24$.

Substitute for $k = 24$ in equation (1) above to obtain

the formula $y = \dfrac{24}{x}$ ....(2)

Step 3. when $x = 2, y = \dfrac{24}{2}$ (Replacing x by 2 in equation (2))

$y = 12$

Therefore, to sort the letters in 2 hours would require 12 people.

Example 4 Assume pressure P is indirectly proportional to volume V

Then P_1 corresponds to V_1 as P_2 corresponds to V_2
Since P is inversely proportional to V,

$P_1:\frac{1}{V_1}$ as $P_2:\frac{1}{V_2}$ and from which

Then $P_1V_1 = P_2V_2$ (product = product)
The above equation is a familiar formula for gases in physics.

Lesson 19 Exercises A (Inverse Proportion)

1. If 2 people can paint a house in 6 days, how many people would paint the house in 3 days?
Answer: 4 people

2. At a speed of 60 miles per hour, a car takes 2 hours to travel from City A to City B. At a speed of 40 miles per hour, how long will it take the car to ravel from City A to City B?
Answer: 3 hours

3. If 6 mail men can sort a given quantity of mail in 4 hours, how many mail men would be needed to sort the same quantity of mail in 2 hours?
Answer: 12 men

4. According to Boyle's law, at constant temperature, the volume of a given mass of a gas is inversely proportional to pressure on the gas. When the volume of a gas is 50 cu. in, the pressure is 10 lb per sq. in. What is the pressure (a) when the volume is 80 cu. in.? ; (b) when the volume is 15 cu. in.?; (c) What is the volume when the pressure is 40 lb per sq. in.?   Answer: (a) $6\frac{1}{4}$ lb per sq. in.; (b) $33\frac{1}{3}$ lb per sq. in; (c) $12\frac{1}{2}$ u. in;

5. The speed of rotation of meshed gears is inversely proportional to the number of teeth. A gear with 12 teeth rotates at 840 rpm. What is the rotation speed of a 16-tooth gear?
Answer: 630 rpm

6. The electric current in a circuit is inversely proportional to the resistance of the circuit. If the current is 22 amps when the resistance is 5 ohms, find the current when the resistance is 10 ohms.   Answer: 11 amps

Lesson 19 Exercises B

1. At constant pressure, the volume V of a perfect gas varies as the absolute temperature T.

   If when the temperature is 2730° Absolute, the volume of one gram of a gas is 1.7 cm^3, find the volume when the Absolute temperature is 4830°.

2. According to Hooke's law, the force required to stretch a spring is directly proportional to the elongation of the spring. If a 20 lb force stretches a spring 6 in., what force will be required to stretch it 8 in.?

3. The cost C of gasoline is proportional to the number of gallons N of gasoline. If 10 gallons of gasoline cost $12, (a) what is the cost of 25 gallons of gasoline? and (b) how many gallons of gasoline can $18 buy?

Answers: **1**. 3.0 cm^3; **2**. $26\frac{2}{3}$ lb; **3**. (a) $30, (b) 15 gallons

Lesson 19 Exercises C

---

1. y varies inversely as x. If when $y = 6$, $x = 2$, determine y when $x = 8$.

2. F is indirectly proportional to the square of D. When $F = 30$, $D = 3$. Find F when $D = 2$.

3. W is indirectly proportional to V. When $W = 15$, $V = 45$ find W when $V = 60$.

Answers: **1.** $\frac{3}{2}$; **2.** $67\frac{1}{2}$; **3.** $11\frac{1}{4}$;

Lesson 19 Exercises D

---

1 According to Boyle's law, at constant temperature, the volume of a given mass of a gas varies inversely as the pressure on the gas. When the volume of a gas is 50 in.^{3}, the pressure is 10 lb per in.^{2}. What is the pressure (a) when the volume is 80 in.^{3}? ; (b) when the volume is15 in.^{3} ?; (c) Find the volume when the pressure is 40 lb per in.^{2} .?

2. If 8 people can complete a piece of work in 40 days, how many people will complete the same piece of work in I0 days? Assume that all the people work at the same rate as one another.

3. At constant temperature, the volume of a given mass of a gas is inversely proportional to the pressure on the gas. When the volume is 150 cu. in,. the pressure is 30 lb per in.^{2}. Find the pressure (a) when the volume is 60 in.^{3} ?; (b) when the volume is 300 in.^{3} .

Ans:**1.** (a) $6\frac{1}{4}$ lb; (b) $33\frac{1}{3}$ lb per in.^{2}; (c) $12\frac{1}{2}$ in.^{3} **2.** 32 people; **3.** (a) 75 lb per in.^{2} .;
(b) 15 lb per in.^{2} .

Lesson 20

Compound Proportion (Variation) Problems

Deriving formulas for any proportion problem, with any number of variables and involving both direct and inverse relationships

Case 1: Joint Proportion (Joint Variation)

If a quantity varies **directly** as **two** or more other quantities (i.e. as the product of two or more other quantities), we call such a proportion **joint proportion or joint variation**.

Example 1: If z varies directly as x and y, then

$$z = kxy \ \text{ or } \ \frac{z}{xy} = k \ \text{(joint variation)}$$

or using subscripted variables, $\dfrac{z_1}{x_1 y_1} = \dfrac{z_2}{x_2 y_2}$ <-- Faster form since you do not have to find k first.

Case 2: Joint Inverse Proportion (Joint Inverse Variation)

The following type of proportion may not be found in current textbooks. However, the author believes that for completeness, this proportion should be included. See the two application examples below. If a quantity varies **indirectly** as **two** or more other quantities (i.e. as the inverse of the product of two or more other quantities), we will call such a proportion inverse **joint proportion or** inverse **joint variation**.

Example   If z varies **indirectly** as x and y, then

$$z = \frac{k}{xy} \ \text{ or } \ xyz = k \ \text{(inverse joint variation)}$$

or using subscripted variables, $z_1 x_1 y_1 = z_2 x_2 y_2$ <-- Faster form

Examples

1. A weight watcher exercising 3 hours per day, 5 times a week lost 10 pounds in 4 months. How many months will it take the weight watcher to lose 10 pounds exercising 2 hours per day, 6 times a week?   **Answer:** 5 months

2. A car traveling at 48 mph and for 6 hours per day covered a certain distance in 10 weeks. How many weeks will it take the car traveling at 60 mph and for 4 hours per day, to cover the same distance?   **Answer:** 12 weeks

Case 3 Combined Proportion (Combined Variation)

If a quantity varies **directly** as one quantity (or as two or more quantities) and **inversely** as another quantity (or other quantities), we call such a variation **combined variation**.

Example 2: If z varies directly as x and inversely as y, then

$$z = k\frac{x}{y} \ \text{or} \ \frac{zy}{x} = k \ \text{(combined variation)}$$

or using subscripted variables, $\dfrac{z_1 y_1}{x_1} = \dfrac{z_2 y_2}{x_2}$ <-- Faster form since you do not have to find k first.

Lesson 20: Compound Proportion (Compound Variation) Problems

Example 3: Assume that pressure P is directly proportional to the temperature T, and inversely proportional to the volume V.

Step 1: Then P_1 corresponds to T_1 directly, and to V_1 inversely as P_2 corresponds to T_2 directly, and to V_2 inversely.

Step 2: Write down P_1 (the first quantity.)

Step 3: For each of the other quantifies, if it is directly proportional to P_1, then it divides P_1 (i.e., write it as a divisor), but if it is inversely proportional to P_1, it multiplies P_1 (that is write it as a factor in the numerator). Similarly, repeat the process beginning with P_2.

$$\text{Then } \frac{P_1 V_1}{T_1} = \frac{P_2 V_2}{T_2}, \text{ a familiar gas law in physics.}$$

Example 4: Assume that A is directly proportional to B and directly proportional to C.

Step 1: Then A_1 corresponds to B_1 directly, and to C_1 directly as A_2 corresponds to B_2 directly, and to C_2 directly.

Step 2: Write down A_1 (the first quantity).

Step 3: For each of the other quantities if it is directly proportional to A_1, then it divides A_1 (That is write it as a divisor) but if it is inversely proportional to A_1, it multiplies A_1 (i.e., write it as a factor in the numerator) . Similarly, on the right side of the equality symbol, repeat the process beginning with A_2.

$$\text{Then } \frac{A_1}{B_1 C_1} = \frac{A_2}{B_2 C_2}$$

Example 5 Assume that the temperature T is directly proportional to the pressure, P, and directly proportional to the volume, V.

Step 1: Then T_1 corresponds to P_1 directly, and to V_1 directly as T_2 corresponds to P_2 directly, and to V_2 directly.

Step 2: Write down T_1 (the first quantity).

Step 3: For each of the other quantifies if it is directly proportional to T_1, then it divides T_1 (i.e., write it as a divisor) but if it is inversely proportional to T_1, it multiplies T_1 (i.e., is write it in the numerator as a factor) Similarly, repeat the process beginning with T_2 .

$$\text{Then } \frac{T_1}{P_1 V_1} = \frac{T_2}{P_2 V_2}$$

By inverting both sides of the above equation, we obtain

$$\frac{P_1 V_1}{T_1} = \frac{P_2 V_2}{T_2}, \text{ the familiar gas law.}$$

Question: Can we obtain the same equation in above example, given the following information?:
Assume volume V is directly proportional to temperature T. and inversely proportional to the pressure P. Then V_1 corresponds to T_1 directly, to P_1 inversely as V_2 corresponds to T_2 directly, and to P_2 inversely.

$$\text{Translating into an equation } \frac{V_1 P_1}{T_1} = \frac{V_2 P_2}{T_2} \text{ or equivalently }, \frac{P_1 V_1}{T_1} = \frac{P_2 V_2}{T_2}. \text{ Answer: Yes.}$$

Example 6: Assume that A is inversely proportional to B and inversely proportional to C.

Then A_1 corresponds to B_1 inversely, and to C_1 inversely as A_2 corresponds to B_2 inversely, and to C_2 inversely.

Then, following the agreement made in the previous examples, we obtain

$A_1 B_1 C_1 = A_2 B_2 C_2$ (**product = product**)

Lesson 20: Compound Proportion (Compound Variation) Problems

Solving Compound Proportion Problems Using Arithmetic

In the previous examples, we covered how to solve compound proportion problems using the variation method. However, we can also solve compound proportion problems using **arithmetic.** We can apply **Method 4** of Lesson 18 (p.66) and Lesson 19 (p.81), (The "More or Less Method")

Example 7. If 8 men can make 20 tables in 6 days, how many men would be required to make 60 tables in 9 days?

Solution:

The number of men required to make 20 tables in 6 days = 8 men

The number of men required to make 60 tables in 9 days = 8 men $\times \dfrac{60 \text{ tables}}{20 \text{ tables}} \times \dfrac{6 \text{ days}}{9 \text{ days}}$

(See p.66 and p.81 for setting up the fractions) $\qquad = 8 \text{ men} \times \dfrac{60}{20} \times \dfrac{6}{9} = 16 \text{ men}$

Lesson 20 Exercises A
(Compound Proportion)

1. If 5 people working 6 hours per day can complete a job in 8 days, how many people working 2 hours per day can complete the same job in 10 days? **Answer:** 12 people

2. If 12 people working 4 hours per day, 3 days per week can complete a job in 4 years, how many people working 9 hours per day, 4 days per week can complete the same job in 2 years? **Answer:** 8 people

3. A weight watcher exercising 3 hours per day, 5 times a week lost 10 pounds in 4 months. How many months will it take the weight watcher to lose 10 pounds exercising 2 hours per day, 6 times a week? **Answer:** 5 months

4. A car traveling at 48 mph and for 6 hours per day covered a certain distance in 10 weeks. How many weeks will it take the car traveling at 60 mph and for 4 hours per day, to cover the same distance? **Answer:** 12 weeks

5. If 10 people working 6 hours per day, 7 days per week earned a certain amount of income each week, how many people working 4 hours per day, 5 days per week would earn the same income in one week? **Answer:** 21 people

6. Six men working 2 days a week can build a fence in 8 weeks. How many men working 4 days a week will build the same fence in 12 weeks. **Answer:** 2 men

7. Six men working 2 days a week can build a fence in 8 weeks. How many weeks will it take 3 men working 4 days a week to build the same fence. **Answer:** 8 weeks

8. If 8 men can make 20 tables in 6 days, how many men would be required to make 60 tables in 9 days? **Answer:** 16 men

Lesson 20 Exercises B

1. Z varies jointly as x and y. If when $x = 2$, $y = 3$, $Z = 16$, what is Z when $x = 5$ and $y = 4$?

2. V is indirectly proportional to P and directly proportional to T. When $V = 30$, $T = 25$, $P = 15$. Find V when $T = 40$, and $P = 3$.

3. F is directly proportional to G and varies inversely as L. When $F = 256$, $G = 36$, and $L = 30$. What is G, when $F = 128$ and $L = 15$?

4. The kinetic energy E of a moving object varies as the mass M. of the object and the square of the velocity V. If when $E = 50$, $M = 4$, $V = 5$, find E when $M = 16$ and $V = 3$.

5. If 12 people working 2 days a week can complete a piece of work in 4 weeks, how many people working 3 days per week can complete the same piece of work in 2 weeks? (Assume that all the people involved work at the same rate as one another)

Answers: **1.** $53\frac{1}{3}$; **2.** 240; **3.** 9; **4.** 72; **5.** 16 people;

EXTRA

Determining the type of relationship, given an equation

Determine how the first variable is related to each of the other variables in each equation.

(a) $\dfrac{P_1 V_1}{T_1} = \dfrac{P_2 V_2}{T_2}$;   (b) $\dfrac{A_1 B_1}{C_1} = \dfrac{A_2 B_2}{C_2}$;   (c) $\dfrac{B_1}{D_1} = \dfrac{B_2}{D_2}$;   (d) $\dfrac{A_1 B_1 C_1}{D_1 E_1} = \dfrac{A_2 B_2 C_2}{D_2 E_2}$

Solution
(a) P is inversely proportional to V and directly proportional to T
(b) A is inversely proportional to B and directly proportional to C
(c) B is directly proportional to D.
(d) A is inversely proportional to B, inversely proportional C, directly proportional to D, and directly proportional E.

Comparison of Sample Formulations of Proportion Problems

Proportion Formulation (subscript notation)	Variation Formulation	Statement of relationship
1. $\dfrac{V_1}{T_1} = \dfrac{V_2}{T_2}$	$\dfrac{V}{T} = k$ or $V = kT$	V is directly proportional to T
2. $P_1 V_1 = P_2 V_2$	$PV = k$ or $P = \dfrac{k}{V}$	P is inversely proportional to V
3. $\dfrac{P_1 V_1}{T_1} = \dfrac{P_2 V_2}{T_2}$	$\dfrac{PV}{T} = k$ or $PV = kT$	P is inversely proportional to V and directly proportional to T.
4. $I_1 R_1 = I_2 R_2$	$IR = k$ or $I = \dfrac{k}{R}$	I is inversely proportional to R
5. $E_1 I_1 = E_2 I_2$	$EI = k$ or $E = \dfrac{k}{I}$	E is inversely proportional to I
6. $\dfrac{E_1}{N_1} = \dfrac{E_2}{N_2}$	$\dfrac{E}{N} = k$ or $E = kN$	E is directly proportional to N
7. $I_1 N_1 = I_2 N_2$	$IN = k$ or $I = \dfrac{k}{N}$	I is inversely proportional to N
8.		
9.		

CHAPTER 7

Lesson 21: **Percent (%) and Inter-conversions**
Lesson 22: **Calculations Involving Percent**
Lesson 23: **More Applications Involving Percent**

Lesson 21
Percent (%) and Inter-conversions

Some interpretations of the percent symbol "%":

1. Over hundred: For example, 20% means $\frac{20}{100}$ (Twenty over hundred or 20 divided by 100)

2. For each hundred: For example, a savings account with a 5% interest rate pays the depositor $5 for each $100 in the account. (Five for each hundred)

3. Hundredths: For example, 20% means $\frac{20}{100}$ or .20   (Twenty hundredths)

4. As a number out of 100: For example, a grade of 80% on a test means a student got 80 points out of 100 points.

Changing a decimal (or any fraction) to percent

Procedure: Multiply by 100%. (i.e., multiply by 100 and attach the percent symbol "%".

To change a decimal to percent, move the decimal point two places to the right and attach the percent symbol. (Note that moving the decimal point two places to the right is equivalent to multiplying by 100)

Example 1 Convert .74 to percent.
Solution   .74 = 74% .

Example 2 Convert .008 to percent.
Solution   .008 = .8%

Example 3 Convert 1 to percent.
 Solution   1 = 100%

Example 4 Convert 12 to percent.
Solution   12. = 1200%

Example 5   Change $\frac{1}{4}$ to percent

Method 1 $\frac{1}{4} \times \frac{100\%}{1} = 25\%$

Method 2 $\frac{1}{4} = .25 = .25 \times 100\% = 25\%$ (Changing to decimal first, and then changing to percent)

Changing a percent to a decimal

Procedure: Divide by 100%. (i.e., drop the percent symbol "%" and divide by 100) or apply the meaning of the percent symbol to change the percent to a decimal fraction and then easily to a decimal. To change a percent to a decimal, drop the percent symbol and move the decimal point two places to the left .

(**Note** that moving the decimal point two places to the left is equivalent to dividing by 100)

Case 1: Example 1 $74\% = .74$ $\left(74\% = \frac{74}{100} = .74\right)$

  Example 2 $.8\% = .008$ (or $.8\% = \frac{.8}{100} = .008$)

  Example 3 $145\% = 1.45$

  Example 4 $14.5\% = .145$

  Example 5 $.145\% = .00145$

Case 2: Example 6 Convert $84\frac{2}{3}\%$ to a decimal

$$84\frac{2}{3}\% = \frac{84\frac{2}{3}}{100}$$

$$= .84\frac{2}{3} \text{ (moving the decimal point two places to the left)}$$

Example 7 Convert $84\frac{2}{3}\%$ to a decimal rounded-off to the nearest **hundredth.**

$84\frac{2}{3}\% = 0.846... \approx 0.85$ $\left(\frac{254}{3}\% = 84.6...\% = .846... \approx .85\right)$

Example 8 Convert $8\frac{1}{4}\%$ to a decimal.

$$8\frac{1}{4}\% = 0.08\frac{1}{4} \text{ <-------- Complex decimal.}$$

However, since $\frac{1}{4} = 0.25$, a terminating decimal,

$8\frac{1}{4}\% = 0.08\frac{1}{4} = 0.0825$ (Also, $8\frac{1}{4}\% = \frac{33}{4}\% = 8.25\% = \frac{8.25}{100} = .0825$.)

Example 9 Convert $4\frac{2}{3}\%$ to a decimal

$$4\frac{2}{3}\% = \frac{4\frac{2}{3}}{100}$$

$$= .04\frac{2}{3} \text{ (moving the decimal point two places to the left and writing a zero to hold place)}$$

To check: Let us convert $.04\frac{2}{3}$ to a percent.

$.04\frac{2}{3} = .04\frac{2}{3} \times 100\%$

$$= 4\frac{2}{3}\% \text{ (moving the decimal point two places to the right and attaching the percent symbol)}$$

Changing a Percent to a Fraction (in its lowest terms)

Example 1. Convert 25% to a fraction in its lowest terms.

Solution: $25\% = \frac{25}{100} = \frac{1}{4}$.

Example 2. Convert 23% to a fraction in its lowest terms.

Solution: $23\% = \frac{23}{100}$.

Example 3. Convert 74% to a fraction in its lowest terms

Solution: $74\% = \frac{74}{100} = \frac{37}{50}$.

Example 4 Convert $4\frac{2}{3}\%$ to a fraction

$$4\frac{2}{3}\% = \frac{4\frac{2}{3}}{100}$$

$$= \frac{14}{300}$$

$$= \frac{7}{150}$$

Example 5 Convert $.4\frac{2}{3}\%$ to a fraction

$$.4\frac{2}{3}\% = \frac{.4\frac{2}{3}}{100}$$

$$= .4\frac{2}{3} \div 100$$

$$= \frac{14}{30} \div 100$$

$$= \frac{14}{3000}$$

$$= \frac{7}{1500}$$

Note: Attaching the % symbol is equivalent to dividing by 100; and dropping the % symbol is equivalent to multiplying by 100.

Lesson 21 Exercises

A. Convert to a decimal: **1.** 23%; **2.** 8.25%; **3.** 0.8%; **4.** 10%; **5.** 10.5%; **6.** 8%

Answers: **1.** 0.23; **2.** 0.0825 **3.** 0.008; **4.** 0.10 or 0.1; **5.** 0.105; **6.** 0.08

B. Convert to a decimal: **1.** $2\frac{2}{3}\%$; **2.** $25\frac{2}{7}\%$; **3.** $4.3\frac{5}{11}\%$; **4.** $0.16\frac{2}{3}\%$; **5.** 8.2%; **6.** $64\frac{3}{11}\%$

Answers: 1. $0.02\frac{2}{3}$; **2.** $0.25\frac{2}{7}$; **3.** $0.043\frac{5}{11}$; **4.** $0.0016\frac{2}{3}$. **5.** 0.082; **6.** $0.64\frac{3}{11}$

C. Convert to a fraction in its lowest terms

1. 24%; **2.** 63%; **3.** 96%; **4.** 8.2%

Answers: 1. $\frac{6}{25}$; **2.** $\frac{63}{100}$; **3.** $\frac{24}{25}$; **4.** $\frac{41}{500}$.

D. Convert to a fraction in its lowest terms

1. $2\frac{2}{3}\%$; **2.** $25\frac{2}{7}\%$; **3.** $4.3\frac{5}{11}\%$; **4.** $0.16\frac{2}{3}\%$.

Answers: 1. $\frac{2}{75}$; **2.** $\frac{177}{700}$; **3.** $\frac{239}{5500}$; **4.** $\frac{1}{600}$.

Lesson 22
Calculations Involving Percent (%)

In calculations involving percent, three main quantities are involved, namely the percentage, the base, and the rate percent. Some authors call the percentage the amount.
In these problems, you are usually given two of these quantities and you are asked to find the third quantity.

*Percentage: This is what is obtained when a percent is taken of a number.

Base: This is the number **of** which a percent is taken.

Rate: This the **percent** that is taken of a number.

There are formulas relating these three quantities:
$$\text{1. percentage} = \text{base} \times \text{rate}\%$$
$$\text{2. base} = \text{percentage} \div \text{rate } \%$$
$$\text{3. rate}\% = \frac{\text{percentage}}{\text{base}} \times 100\% \qquad (\text{i.e. rate}\% = (\text{the ratio of percentage to base}) \times 100\%)$$

You do not need to memorize the first formula, provided you note that "**of**" implies multiply.
Memorize the second and the third formulas (even though some of the methods discussed below do not need the recall of these formulas). *Some authors call the percentage the **amount** and suggest the proportion

$$\frac{r}{100} = \frac{\text{Percentage}}{\text{base}} = \frac{r}{100} = \frac{A}{B}, \text{ where } r = \text{rate}, A = \text{Amount and } B = \text{Base}.$$

Finding the Percentage

Example 1   Find 20% of 72   (i.e. **Finding the Percentage**)

(Note: "%" means over 100. Example: $20\% = \frac{20}{100}$

20% of 72   ("of" means multiply)

Step 1: $= \frac{20}{100} \times \frac{72}{1}$ <---------Simplify this by any of the methods
discussed below.

Step 2:

Method 1: $\frac{20}{100} \times \frac{72}{1} = .20 \times 72 = 14.40 = 14.4$ (Using decimals)

Method 2: $\frac{20}{100} \times \frac{72}{1} = \frac{20 \times 72}{100} = \frac{1440}{100} = 14.40$ or 14.4 (Multiplying   numerators
and dividing by 100)

Method 3: $\overset{1}{\cancel{\frac{20}{100}}}_{5} \times \frac{72}{1} = \frac{72}{5} = 14\frac{2}{5}$ or 14.4   (using cancellation)

The most convenient method will depend on the type of numbers involved, and whether we want the answer as a decimal, as a fraction, or as a mixed number.
For example, if there are common factors in the numerator and the denominator, cancellation, (Method 3) may be more convenient; but if there are no common factors , use Method 1 or Method 2.
 In any case, it is a good practice to set up the problem as in Step 1. The next example will show t he usefulness of setting up the problem before proceeding to simplify.

Example 2 Find $4\frac{2}{7}\%$ of 28000 .

Step 1 : Translating: $4\frac{2}{7}\%$ of 28000

$$= \frac{4\frac{2}{7}}{100} \times \frac{28000}{1}$$

Step 2: $= \frac{30}{700} \times \frac{28000}{1}$ <-----------Simplify this by any method.

The easiest method is by cancellation, since there are common factors in both the numerator and the denominator.

$$\frac{30}{\cancel{700}^{1}} \times \frac{\cancel{28000}^{40}}{1} = 30 \times 40 = 1200$$

Note that steps 1 and 2 are important. Do not round off $4\frac{2}{7}\%$ as .04% because you will not get the exact answer. If the rate % were say, 70%, then you could immediately write 70% = .70 and then similarly write $70\frac{1}{4}\%$ = .7025 ; but better, $70\frac{1}{4}\%$ = $\frac{281}{400}$.

Finding the Base (Finding the Original Number) 94

Note that the next problem is different from the last two examples both in the wording of the problem and how we solve it. In the last two problems, we multiplied. In the next problem; we will divide, but, we must note which number is the divisor.

Example If 20% of a number is 64, what is the number?

There are a number of methods for solving this problem. We will discuss five methods, which include algebraic methods. You may skip the algebraic methods if you do not have the algebraic background to allow you to follow the procedure. Later, in Chapter 7 (percent problems) we will repeat the algebraic method.

Method 1:   Using Formula

Base = Percentage ÷ Rate %

In the above problem, 64 is the percentage. (The percentage is sometimes called the " is number". It is the number that (usually) immediately follows or precedes the word "is " in the word problem.

$$20\% \text{ is the rate}\%$$

$$\text{base} = \text{percentage} \div \text{rate} \%$$

$$= 64 \div 20\%$$

$$= \frac{64}{1} \div \frac{20}{100} \quad \text{<-----you can simplify this by approach 1 or 2 below.}$$

Approach 1. $\dfrac{64}{1} \times \dfrac{\cancel{100}^{\,5}}{\cancel{20}} = 64 \times 5 = 320$

Approach 2. $64 \div .20$ (by long division) $20\overline{)6400}^{\,320}$

The number is 320.

Method 2   Using Algebra

Let the number be x.

Then, "20% of the number is 64" translates to $\dfrac{20x}{100} = 64$

i.e. $\dfrac{20x}{100} = 64$

Solve for x: $\dfrac{(\cancel{100})^{1}}{(\cancel{20})_{1}} \dfrac{\cancel{20}x^{1}}{\cancel{100}} = 64(\cancel{100})^{5} \over (\cancel{20})_{1}$

(or $20x = 100 \times 64$

$x = 320$ (or $x = \dfrac{100 \times 64}{20} = 320$)

Again, the number is 320 .

Note also that since 20% of the number is 64, the number must be greater than 64.

Method 3   If 20% of the number = 64

then 1% of the number $= \dfrac{64}{20}$

and 100% of the number $= \dfrac{64}{20} \times \dfrac{100}{1}$

$= 320.$

Method 4   "**Ratio Method**" **(Arithmetic)**
(This method follows from method 3.)

If 20% of a number = 64

then, 100% of the number $= \dfrac{64}{1} \times \dfrac{\cancel{100\%}^{5}}{\cancel{20\%}}$

$= 320$

In method 4, we used a very useful principle which states that " If more, the smaller divides, and if less, the larger divides". That is, since we expect 100% of the number to be greater than 64, in forming the fraction involving 20% and 100%, the smaller of 20% and 100% is the divisor (smaller divides), and hence we used the fraction,

$\dfrac{100\%}{20\%}$ <----- " smaller divides" (We multiplied by an improper fraction)

However, if we had expected the number (answer) to be less than 64, we would have used the

fraction $\dfrac{20\%}{100\%}$ <----- "larger divides" (i.e., we would have multiplied by a proper fraction)

Method 5   **Using Proportion**

20% is to 64 as 100% is to x, where 100% of the
original number (the base) is x.
Translating the proportion,

$$\dfrac{20\%}{64} = \dfrac{100\%}{x} \qquad\qquad (\text{or } \dfrac{.20}{64} = \dfrac{1}{x})$$

Solve for x: 20% x= 100% (64)   (or $.20x = 64$)

$$x = \dfrac{100\%}{20\%}(64) \qquad\qquad (\text{or } x = \dfrac{64}{.20}$$
$$x = 320.$$

Note: Methods 3, 4 and 5 are basically the same.

Note that 20% as a fraction is $\frac{20}{100} = \frac{1}{5}$.

Therefore, in the above problem, instead of asking "if 20% of a number is 64, what is the number?", we could have asked "if $\frac{1}{5}$ of a number is 64, what is the number? "

Example   If $\frac{1}{5}$ of a number is 64, what is the number? We can solve this by any of the methods discussed above. Let us use the algebraic method.

Solving Algebraically:

Let the number be x.

Then $\frac{x}{5} = \frac{64}{1}$ (Note that $\frac{1}{5}x = \frac{x}{5}$)

$(5)\frac{x}{5} = \frac{64}{1}(5)$ or $x \times 1 = 5 \times 64$ (by cross-multiplication)

$x = 320$ or $x = 320$.

The number is 320.

(See also Method 1 of the preceding example)

Go over the last two problems, and note the differences between how the questions are worded and how they are solved. You **must remember** how each problem is worded and how to proceed to solve it.

Note also that the previous example was a direct proportion problem. Direct proportion involves a relationship between two quantities whereby as one quantity **increases** the other quantity also **increases.** See p.64 for more examples on direct proportion and also examples on inverse (indirect) proportion whereby as one quantity **increases** the other quantity **decreases** and vice versa.

Example   Mary spends 20% of her weekly income on food. If she spends $64 on food every week, what is her weekly income?

Solution : We could reword this problem in the familiar form as " If 20% of a number is 64, what is the number?"; and then use exactly the same method as in the example, above.

Answer: $320 (numerically, the same answer as for the last example).

Finding the Rate %

Example 1 What rate % of 24 is 15?

Method 1

$$\text{By formula: Rate\%} = \frac{\text{Percentage}}{\text{Base}} \times 100\%$$

In this problem , the percentage is 15. The percentage is sometimes referred to as the " is number". It is the number that follows or precedes the word "is" if the problem is worded in the above form. The base is 24. The base is sometimes referred to as the "of number". It is the number that (usually) follows the word "of" if the problem is worded in the above form.

$$\text{Then, rate \%} = \frac{15}{24} \times \frac{100\%}{1}$$
$$= \frac{5\cancel{15}}{\cancel{24}_8} \times \frac{100\%}{1}$$
$$= .625 \times 100\%$$
$$= 62.5\% \text{ or } 62\tfrac{1}{2}\%$$

Method 2

Step 1: Form a fraction using the mnemonic device " $\dfrac{\text{is number}}{\text{of number}}$ "

Then , we obtain $\dfrac{15}{24}$

Step 2: Change $\dfrac{15}{24}$ to percent by multiplying by 100 and attaching the '"%" symbol

$$\frac{5\cancel{15}}{\cancel{24}_8} \times \frac{100\%}{1} = \frac{500\%}{8}$$
$$= 62\tfrac{1}{2}\% \text{ or } 62.5\%$$

Method 3 Using algebra <--You may skip this method if you do not have the algebraic background to allow you to follow the procedure. Later, in Chapter 7, we will repeat this method.

Let $x\%$ be the rate %. We will write an equation in terms of x, and solve for x.
From the wording, "$x\%$ of 24 is 15" translates to:

$$\frac{x}{100} \times \frac{24}{1} = 15$$

$$\frac{24x}{100} = \frac{15}{1}$$
$24x = 100 \times 15$ (by cross-multiplication or by multiplying both sides of the equation by 100)
$x = \dfrac{100 \times 15}{24}$ (by dividing both sides of the equation by 24)
$x = 62\tfrac{1}{2}$

$\therefore$ the required rate $= 62\tfrac{1}{2}\%$.

Example 2   What rate % of 18 is 44?

Method 1   base = 18 (the "of" number)
                         percentage = 44 (the "is" number)

$$\text{rate } \% = \frac{\text{percentage}}{\text{base}} \times 100\%$$

$$= \frac{\overset{22}{\cancel{44}}}{\underset{9}{\cancel{18}}} \times \frac{100\,\%}{1}$$

∴ the required rate is $244\frac{4}{9}\%$

Note in the above problem that, the larger number does NOT have to be in the denominator. The base (the "of" number) must always be in the denominator. Therefore, in forming the fraction, ignore the relative sizes of the numbers.

Method 2 (Using algebra)
Let the required rate be $x\%$.
Then, "$x\%$ of 18 is 44" translates to:

$$\frac{x}{100} \times \frac{18}{1} = 44$$

$$\frac{18x}{100} = \frac{44}{1}$$

$18x = 100 \times 44$ (by cross-multiplication or by multiplying both sides of the equation by 100)

$x = \frac{100 \times 44}{18}$ (by dividing both sides of the equation by 18)

$x = 244\frac{4}{9}$

∴ the required rate is $244\frac{4}{9}\%$

Example 3   A family's annual income last year was \$20,000. This year, the income
              is \$33 ,000. What is the percent increase in the annual income?

Solution

Step 1: The increase in income = \$33,000 - \$20,000
                            = \$13,000

Step 2: The percent increase in income $= \frac{13000}{20000} \times 100\%$ (Finding the rate percent)
                            = 65%

Note: In Step 2 , the question could have been posed as: What percent of 20,000 is 13.000 ?

Example 4   On a class test , out of 20 questions, a student answered
               16 questions correctly. What was the student's grade in percent?

Step 1:   Fraction of questions answered correctly $= \frac{16}{20}$

Step 2: Change $\frac{16}{20}$ to percent by multiplying by 100 and attaching the % symbol.

$$\frac{16}{20} \times 100\% = 80\%$$

The student's grade was 80%.

Lesson 22 Exercises

A. (*a*) Find 80% of 25;　　　(*b*) Find 23% of 60;　　　(*c*) Find $5\frac{2}{9}$% of 1800

Answers　(*a*)　20;　(*b*)　13.8 or $13\frac{4}{5}$;　(*c*) 94

B. 1. If 20% of a number is 72, what is the number?

2. 53 is 25% of what number?

3. 16% of what number is 82?

4. If 25% of a number is 140, what is the number?

5. If 120% of a number is 103.2, what is the number?

Answers **1.** 360 ; **2.** 212; **3.** 512.5 or $512\frac{1}{2}$; **4.** 560 ;　**5.** 86.

C. 1. The monthly rent for Maria's apartment is $800. If Maria spends 25% of her monthly income on this rent, what is her monthly income?

2. A homeowner borrowed money from a bank at the interest rate of 12% per year. If the homeowner pays the bank an interest of $6,000 per year, how much money did the homeowner borrow?

3. If $3\frac{2}{7}$ of a number is 46, What is the number?

Answer: 1. $3,200;　**2.** $50,000;　**3.** 14.

D. 1. What rate percent of 32 is 12?

2. What rate% of 20 is 80?

3. On a math test, there were 40 questions. James answered 32 questions correctly. What was his grade in percent?

Answers　**1.** 37.5% or $37\frac{1}{2}$%; **2.** 400% ; **3.** 80%

Lesson 23

More Applications Involving Percent:
Discount , Salary Change and Sales Tax Problems

Applications of Base Finding and Percentage Finding

In these applications, the questions are **not** worded in forms such as "find 20% of a number; "if 30% of a number is 45, what is the number?". A good approach is to reword the problem in any of these familiar forms and then proceed accordingly.

Note 1. The fraction involved in the problem may very likely be the **rate** (but it may **not** be the rate).

Note for example that $\frac{3}{5} = 60\%$

2. The quantity following the word "of" may very likely be the **base**.

If you know any two of the three quantities, then third quantity is easily deduced.

Example 1   3 out of 5 students at a certain college study Biology.

(a) If 600 students at this college study Biology, how many students are there at this college?

(b) If one were to collect a sample of 200 students at this college, how many of these students would study Biology?

Solution:

There are a number of approaches for solving this problem. We will cover two methods.

Note that 3 out of 5 means $\frac{3}{5}$

Part (a): We can reword this part of the question as: if $\frac{3}{5}$ of a number is 600, what is the number?

Method 1   Let the number be x

Then $\frac{3x}{5} = 600$

$3x = 3000$

$x = 1000$

There are 1000 students at this college.

Method 2 (see also p. 94)   Divide 600 by $\frac{3}{5}$

$$600 \div \frac{3}{5} = \frac{600}{1} \times \frac{5}{3}$$
$$= 1000$$

Part (b)   We can reword this part of the question as: Find $\frac{3}{5}$ of 200.

$$\frac{3}{5} \text{ of } 200 = \frac{3}{5} \times \frac{200}{1} = 120$$

That is, of a sample of 200 students, 120 students would study Biology.

Example 2

Note that since $\frac{3}{5} = 60\%$, the above problem could have been posed as 60% of students at a certain college study Biology.

(a) If 600 students study Biology, how many students are there?

(b) If one were to collect a sample of 200 students, how many would study Biology?
Solution: Proceed exactly as in Example 1 above.

Example 3 At a certain college, 53% of students registered for chemistry, and 24% registered for Biology. If there are 500 students at this college, how many students did not register for Chemistry or Biology?

Solution: We will use two methods to solve this problem.

Method 1
Step 1: Percent of students registering for Chemistry or Biology is
$(53\% + 24\%) = 77\%$

Step 2: Percent of students **not** registering for Chemistry or Biology is
$100\% - 77\% = 23\%$

Step 3: Number of students who did not register for Chemistry or Biology is

$$23\% \text{ of } 500$$
$$= \frac{23}{100} \times 500$$
$$= 115.$$

Method 2
Step 1: Percent of students registering for Chemistry or Biology is
$(53\% + 24\%) = 77\%$

Step 2 : Find 77% of 500 and subtract the result from 500.

$$\text{i.e. } \frac{77}{100} \times 500 = 385$$

Then, the number of students not registering for Chemistry or Biology is

Step 3: $500 - 385 = 115$

By either Method 1 or Method 2, the number of students not registering for Chemistry or Biology is 115.

Discount Problem

Example   A bag originally sold for $85.00

The selling price was reduced by 30%
What is the new selling price?

Method 1

Step 1: Find 30% of $85.00

i.e. $\frac{30}{100} \times \frac{85}{1} = 25.50$ or $\begin{array}{r} 85 \\ \times .30 \\ \hline 25.50 \end{array}$

Step 2: New Price = $85.00 - $25.50
                       = $59.50

Method 2   Assuming a 100% rate % for the original price,

Step 1:   Subtract 30% from 100%
                i.e. 100% - 30% = 70%

Step 2 :   Find 70% of $85.00

i.e. $\frac{70}{100} \times \frac{85}{1} = \$59.50$ or $\begin{array}{r} 85 \\ \times .70 \\ \hline 59.50 \end{array}$

Salary Change Problem

Example

A year ago, Mary's annual salary was $45,000. This year she received a 15% raise.
What is her new salary?

The approach in solving the problem is similar to that of the above discount
problem, except that in this case, we add any change (in salary).

Method 1   Step 1:   Find 15% of 45,000
                $\frac{15}{100} \times \frac{45000}{1} = 6,750$
                The increase (raise) in salary is $6,750

Step 2: New Salary = Original Salary + Increase in salary
                     = $45,000 + $6,750
                     = $51,750

Method 2

Step 1:   Add 15% to 100%.
                15% + 100 % = 115%

Step 2:   Find 115% of 45,00
                $\frac{115}{100} \times \frac{45000}{1} = \$51,750$

Sales Tax Problem

Example   A book sells for $50.00
and the sales tax is 8%.
How much does the purchaser pay?

Method 1

Step 1: Find 8% of $50

$$\frac{8}{100} \times \frac{50}{1} = 4 \qquad \text{or} \qquad \begin{array}{r} 50 \\ \times .08 \\ \hline 4.00 \end{array}$$

the sales tax is $4.00

Step 2:   Total cost = $50.00 + $4.00
=$54.00
∴ The purchaser pays $54.00

Method 2   Step 1: Add 8% to 100%
i.e.   8% + 100% = 108%

Step 2: Find 108% of $50.00
$$\frac{108}{100} \times \frac{50}{1} = \$54.00$$

Lesson 23 Exercises

1. A new math textbook sells for $32.00. However, a used edition of this book is sold at a 6% discount. What is the selling price of the used edition?

Answer $30.08

2, Last year , the president of a corporation was earning $160,000 per year. This year, because of financial problems, the annual salary is to be reduced to $140,000.
What is the percent decrease in salary?

Answer 12.5% or $12\frac{1}{2}\%$

3. A book sells for $65.00 and the sales tax is 8.25%.
What is the total price?

Answer $70.36

CHAPTER 8
Miscellaneous Topics

Lesson 24
Averages, Duration, Profit and Loss,

Finding Averages

Example 1   Find the average of 6, 8, 11, and 15.

We must **always remember** that in finding an average, we add all the quantities and divide by the number of quantities.

$$
\begin{array}{r}
6 \\
8 \\
11 \\
\underline{15} \\
40
\end{array}
$$

The number of quantities is 4.
We divide 40 by 4
Then, $\frac{40}{4} = 10$
The average is 10

Example 2   During the semester, James scored 85, 70, 75 on the first three tests. He wants his class average for the term to be 80. What is the score he needs on the next and last test ?

Solution   We cover three methods.

Method 1: Arithmetic approach.

Step 1:   For an average of 80 on 4 tests, total score must be $80 \times 4 = 320$

Step 2:   Total score on the first three tests is $75 + 70 + 75 = 230$

Step 3:   Score needed on the 4th test to make total score 320 is
$320 - 230 = 90$

∴   James needs a score of 90 on the next test.

Method 2: Layperson's approach
Step 1:

1st test	2nd est	3rd test	4th test
85	**70**	**75**	**?**
Here, there are an extra 5 points to spare	Here, we need 10 points	Here, we need 5 points	Here, we need 80 points

Step 2: If we add the 5 extra points from the first test to the 3rd test, we would have

1st test	2nd test	3rd test	4th test
80	**70**	**80**	**?**
This is O.K	Here, we need 10 more points	This is O.K	Here, we need 80 points

Since there is only one more test to be taken, James must make up the 10 points needed for the 2nd test, and obtain 80 points needed for the 4th test , all on the last test.

Hence, James needs a score of $10 + 80 = 90$ on the last test.

Method 3: Using algebra.

Let x be the next and last score.

Then, the total score for the four tests will be $85 + 70 + 75 + x$

$$\text{Average score} = \frac{\text{Total score}}{\text{Number of tests}} = \frac{85 + 70 + 75 + x}{4}$$

Now we want this average score to be 80

$$\frac{85 + 70 + 75 + x}{4} = 80 \qquad \text{(Solve for } x\text{)}$$

$$\frac{230 + x}{4} = 80 \qquad (85+70+75 = 230)$$

$$\frac{4(230 + x)}{4} = 80(4)$$

$$230 + x = 320$$
$$x = 90$$

$\therefore$ James needs a score of 90 on the next and last test.

Travel time, and Duration Problems

Subtract : 4 hrs. 10 mins
2 hrs. 40 mins

Solution: Note that 1 hr.= 60 mins)..Borrow 1 hr from 4 hr, change it to 60 minutes and add it to 10 mins

$$\begin{array}{l} 3 \quad (60 + 10) = 70 ----- > \quad 3\text{hrs} \qquad 70 \text{ mins} \\ \cancel{4\text{hrs}} \quad \cancel{10 \text{ mins}} \\ - \; 2\text{hrs} \quad 40 \text{ mins} ------- > \quad \underline{-2\text{hrs} \qquad 40 \text{ mins}} \\ \qquad\qquad\qquad\qquad\qquad\qquad\qquad 1\text{hr} \qquad 30 \text{ mins} \end{array}$$

As a word problem, the above problem could have been posed as:

(a) A student left school at 2:40 pm and arrived home at 4:10 pm. How long did it take the student to get home?   or

(b) A movie began at 2: 40 pm and ended at 4:10 pm., how long was the movie? For (a) or (b), the answer is 1 hr. 30 mins

Profit or Loss Problems

Example

Tickets to a concert cost $0.45 each .
Answer the following questions.

(a) How many tickets must be sold in order to collect $47.25

(b)   How much money is collected if 135 tickets are sold?

(c) If expenses for organizing the concert is $120.00   and a 1000 tickets are sold, how much profit does the organizer of the concert make?

Solution: (a) Divide the amount collected by the cost of each ticket . $\dfrac{\$47.25}{\$.45}$.

$$.45\overline{)47.25}$$

$$\downarrow$$

$$105$$
$$45\overline{)4725}$$

105 tickets must be sold.

(To check multiply 105 by .45 to obtain $47.25)

(b)   Multiply the cost of each ticket by the number of tickets sold.

Then, $\$.45 \times 135 = 60.75$

or

$$\begin{array}{r} 135 \\ \times .45 \\ \hline 675 \\ 540 \\ \hline 60.75 \end{array}$$

$\therefore$ $60.75 are collected.

(c) Profit = Amount collected for tickets -   Expenses
 = (1000 × $.45) – $120
 = $450 - $120
 = $330

The organizer makes a profit of $330.

*A Problem Worth Knowing (Time and Work Problem)

Working alone, Maria can paint an apartment in 4 hours, while James can paint the same apartment in 6 hours, working alone. How long it will take Maria and James to paint the same apartment if they both paint together?

Solution

In 1 hour, Maria will paint $\frac{1}{4}$ of the apartment

In 1 hour, James will paint $\frac{1}{6}$ of the apartment

In 1 hour, Maria and James (working together) will paint $\frac{1}{4} + \frac{1}{6} = \frac{5}{12}$ of the apartment.

Now, If $\frac{5}{12}$ of the apartment is painted in 1 hour,

then, the whole apartment will be painted in 1 hour $\div \frac{5}{12}$

$$= 1 \text{ hour} \times \frac{12}{5} = 2.4 \text{ hours}$$

*Note: In elementary algebra, this problem will be solved using algebra.

Lesson 24 Exercises

1. A student's grades for class tests were 73, 64, 0, 85, and 78. What was the student's average class grade ?

Answer 60

2. Betty has $850, $700, $750 in three different bank accounts. This morning, she has decided to open a fourth bank account, today. What is the total amount of money Betty needs to deposit so that at the end of the day, she would have exactly $800 in each of the four bank accounts?

Answer: $900

3. Today, Hilda worked from 7:45 a.m. to 11:15 a.m. For how long did she work today?

Answer: 3 hrs. 30 mins

4. A trader bought 45 bags at $4.20 each and sold each bag for $5.80.
   (a) What was the profit made ?   b) What was the profit percent ?

Answer (a) $72; (b) 38.1%

Lesson 25

Areas and Perimeters; Bar, Line and Circle (Pie) Graphs;

Areas and Perimeters of Rectangles

Example 1 Find the area of the rectangle shown below.

Solution: Area of a rectangle = length × altitude (or width)
$$\text{Area} = (15)(6)$$
$$= 90 \text{ square units}$$

(Area, $A = L \times W$, where L is the length and W is the altitude or the width)

Example 2   In Example 1 above, find the perimeter of the figure.

Solution: The perimeter is the distance around the figure.

The perimeter is $(15 + 6 + 15 + 6) = 42$ units.

(Perimeter, $P = L + W + L + W = 2L + 2W$

Example 3   How many yards of fence are needed to enclose a rectangular plot
of land of dimensions 15 yards by 6 yards?

Answer: 42 yards. (see Example 2 above)

Note that yards of fence is this problem implies the perimeter of the rectangular plot.

Example 4   The dimensions of the floor of a room are 15 yards by 12 yards.
If carpet costs $4.00 per sq. yard, what is the cost of carpet needed
to cover the floor of this room?

Solution: Area of the floor = (15×12) sq. yards

Cost of carpet needed = $((15 \times 12) \times 4)$

$$= \$720.00$$

Note that area is a measure of the **size** of a two-dimensional **surface,** but perimeter is a
measure of the total distance (**length**) **around** the figure.

Example 5 Find the area of the figure below.

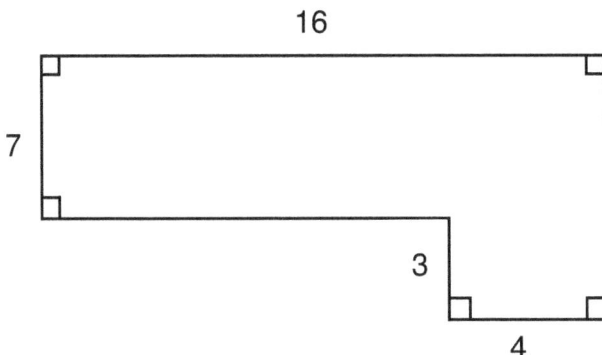

16

7

3

4

Solution: We will break the figure into two rectangles ; find the area of each rectangle (using the
Method 1 area formula for a rectangle) and then add the individual areas to obtain the total area
of the figure.

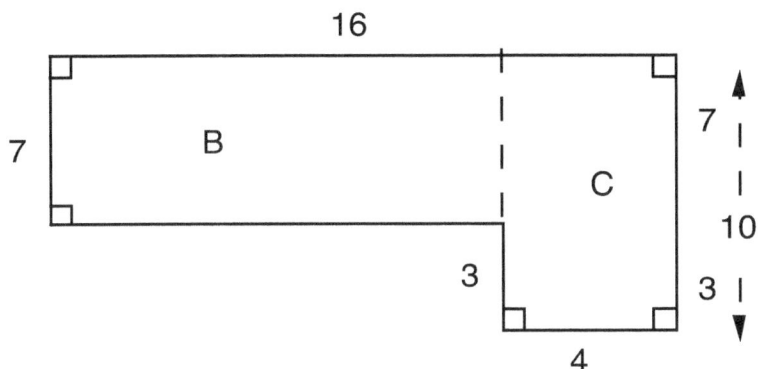

Area of rectangle B = 7(12)   Scrapwork: (16 - 4) =12
              = 84 square units
Area of rectangle C   = 4(10)
              = 40 square units
Total area = area of rectangle B + area of rectangle C
       = 84 sq. units + 40 sq. units
       = 124 square units

Method 2

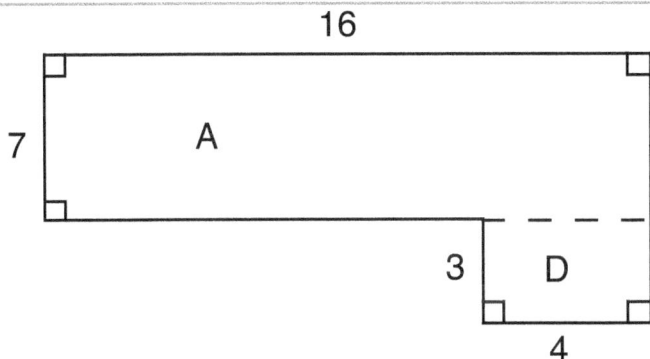

Area of rectangle A = 7(16)
              = 112 square units
Area of rectangle D   = 3(4)
              = 12 square units
Total area
  = area of rectangle A + area of rectangle D
  = 112 sq. units + 12 sq. units
  = 124 square units

Bar Graphs
Example

The bar graph below shows the number of students studying chemistry at certain school during the four years 1980 - 1983.

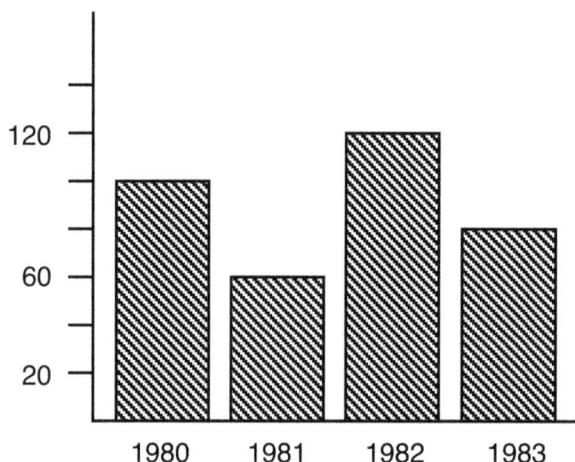

(a) How many students (approximately) studied chemistry in the two years 1980 and 1982?

(b) How many students (approximately) studied chemistry in the years 1980 - 1983?

Solution:

To determine the number of students for any particular year, pick the year (labeled horizontally) and move up vertically to the top of the box for that year. The corresponding reading (labeled) vertically on the left is the number of students for that year.

then, (a) For the year 1980 we read 100 students (approximately)
for the year 1982, we read 120 students (approximately)

The total number of students for the years 1980 and 1982 is (100 + 120) = 220. Therefore, 220 students studied chemistry in the two years 1980 and 1982.

(b) Similarly, for 1980, we read 100 students
1981, we read 60 students
1982, we read 120 students
1983, we read 80 students

Total number of students who studied
chemistry during 1980 - 1983   = 360 students

Line Graphs

We will reformulate the bar graph example (last example) in the form of a line graph but with respect to the population of a small village.

Example   The line graph below shows the population of a small village during the
years 1980- 1983.
What is the difference between the populations in 1983 and in 1981?

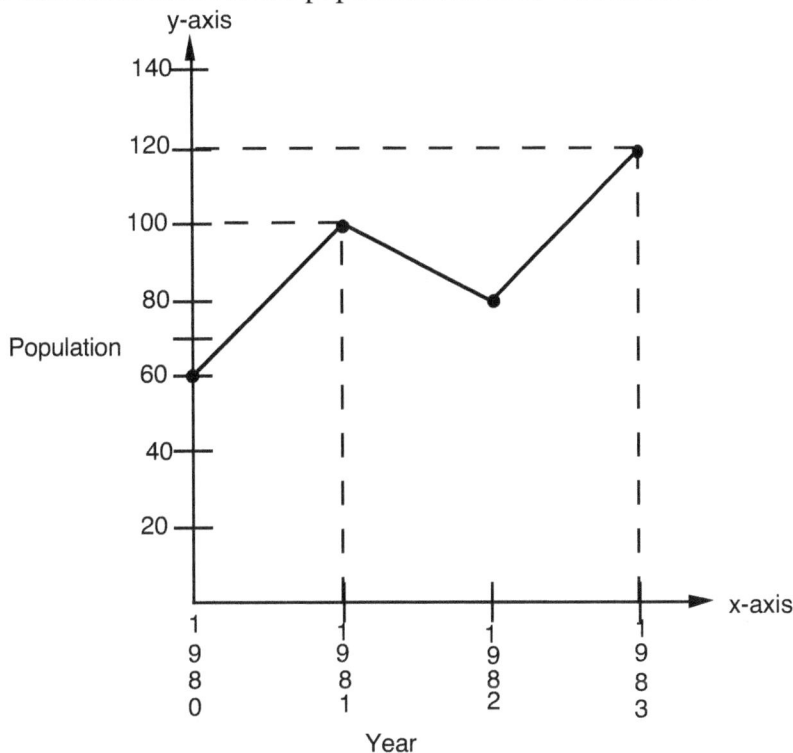

Solution : Here, we consider each point on the line as an ordered pair, where the x-coordinate is the year and y-coordinate the population.
To determine the population for any year, locate on the x-axis, the point corresponding to that year; then, from this point, move up **vertically** until the line graph is met (intersected) .
From this intersection point, move horizontally to the left to the y-axis (the vertical axis) to read the corresponding population for that year.

For the year 1981, we read approximately 100 people.
For the year 1983, we read approximately 120 people.

The difference between the 1983 population and that of 1981 = 120 - 100
i.e. 20 people

Note above that for the year 1980 ,we do not have to move horizontally, since the point of intersection happens to be on the y-axis. For 1980, we read 60.

Circle or Pie Graphs

Example

The pie graph below shows the percent distribution of grades for a math test.

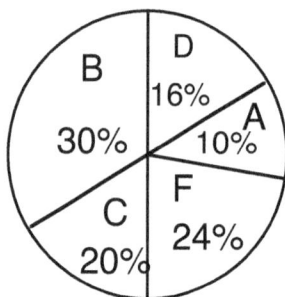

(*a*)   What percent of students received grades of C or better?

Solution: Receiving a grade of C or better means the student received either
a C, a B or an A.

From the graph, % of C grade   =   20 %

% of B grade   =   30 %

% of A grade   =_   10 %

(a) % of students receiving grades C   or better   =   60 %

(*b*)   If there are 90 students in this class, how may students received grades of C or better?

Solution: Percent receiving grades of C or better is 60 % (from part (a)).

∴ Number of students receiving grades of C or better is 60 % of 90

i.e. $\frac{60}{100} \times \frac{90}{1} = 54$

Lesson 25 Exercises

A. 1. Find the area of the figure below.　　　**2.** Find the perimeter of the figure.

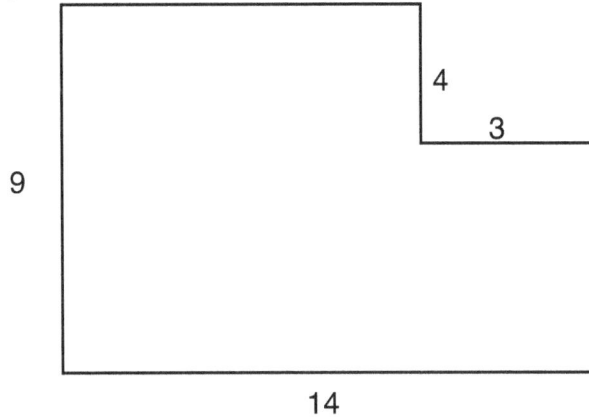

4

3

9

14

Answers: **1.** The area = 114 sq. units; **2.** The perimeter = 46 units

B. 1.. Find the area of the figure below.　　**2.** Find the perimeter of the figure below.

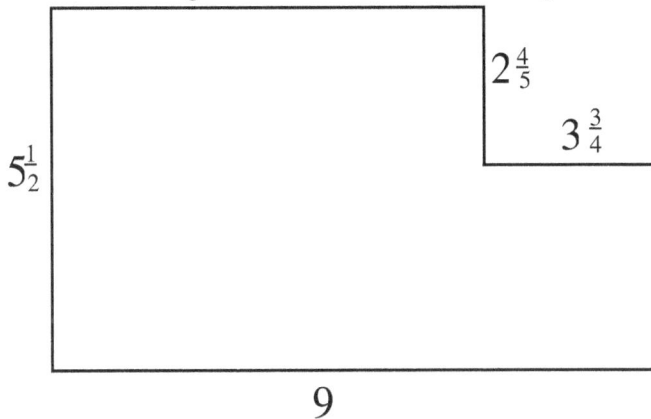

$2\frac{4}{5}$

$3\frac{3}{4}$

$5\frac{1}{2}$

9

Answers: **1.** The area = 39 sq. units;　　**2.** The perimeter = 29 units

C. In the bar graph below, find the difference between the number of students who studied chemistry in 1982 and in 1983.

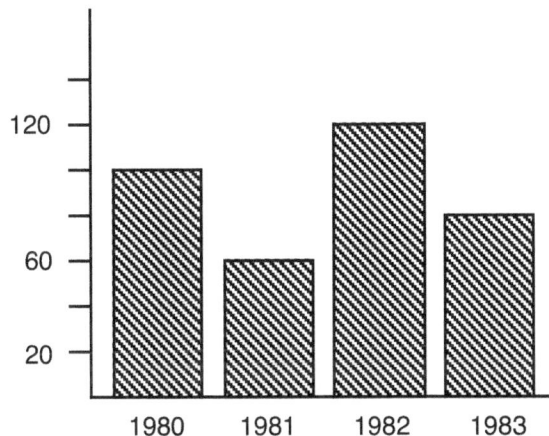

120

60

20

1980　1981　1982　1983

Answer : 40 students

D. The line graph below shows the population of a small village during the years 1980- 1983
1. Find the ratio of the 1982 population to the 1981 population.

2. Find the product of the 1983 population and the 1980 population

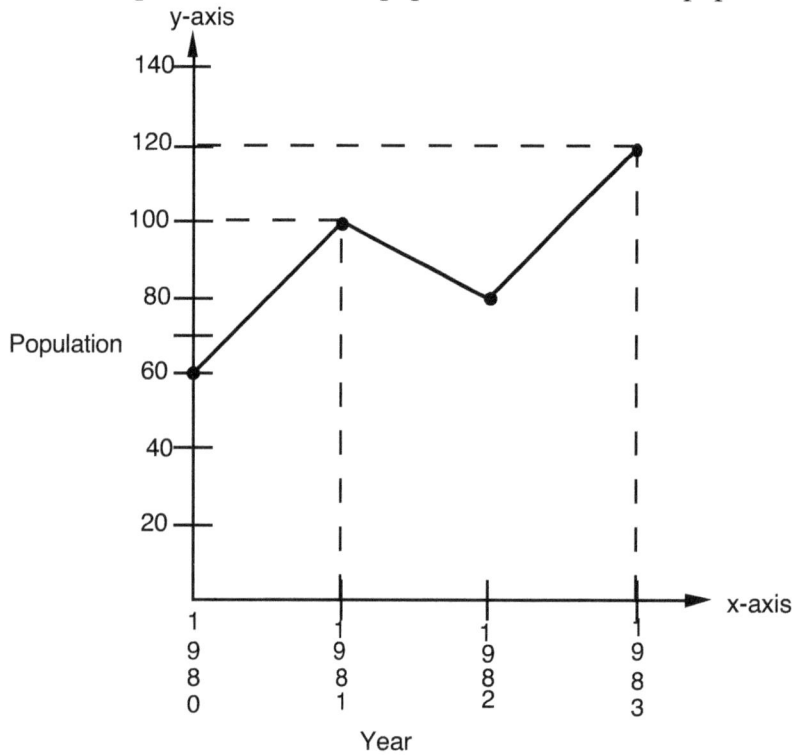

Answers: 1. $4:5$ or $\frac{4}{5}$; **2.** 7200

E The pie graph below shows the percent distribution of grades for a math test.
1. Determine the percent of students with grades of C and D.
2. If there are 25 students in the math class, how many students received grades of C and D?

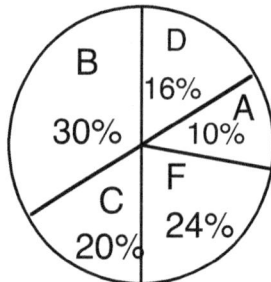

Answers: **1.** 36% **2.** 9

Lesson 26
Scientific Notation

Preliminaries:

Powers of 10

Examples:
$$10^1 = 10$$
$$10^2 = 100$$
$$10^3 = 1000$$
$$10^4 = 10,000$$
} positive integral powers of 10.

Also,
$$10^{-1} = \tfrac{1}{10}$$
$$10^{-2} = \tfrac{1}{100}$$
$$10^{-3} = \tfrac{1}{1000}$$
$$10^{-4} = \tfrac{1}{10000}$$
} negative integral powers of 10.

The inverse of multiplying by 1000 is dividing by 1000 (or multiplying by $\tfrac{1}{1000}$)

The inverse of multiplying by 10^3 is multiplying by 10^{-3} (Since $10^3 = 1000$ and $10^{-3} = \tfrac{1}{1000}$)

Definition: A number is in scientific notation if it is written as the product of a number between 1 and 10 and an integral (integer) power of 10.

Examples The following are in scientific notation:

(a) 2.45×10^4; (b) 5.6×10^{-3} ; (c) 6.023×10^{23}

Procedure for writing a given number in scientific notation:

To write a number in scientific notation, place the decimal point immediately after the first non-zero digit, and multiply by an appropriate power of 10 so that the value of the original number remains unchanged.

Example Write the following in scientific notation:
(a)   6434 ;   (b) .27898   (c) 830,000 ;   (d) .004039

Solution

(a) $6434. = \boxed{6.434 \times 10^3}$ (Move the decimal point three places to the left (i.e., $\times 10^{-3}$) and on the right-hand side of the equation, do the opposite of $\times 10^{-3}$, which is $\times 10^3$.)

$\times 10^{-3}$

Check: $6.434 \times 10^3 = 6.434 \times 1000 = 6434$ (It is good practice to always check the result.)

(b) $.27898 = \boxed{2.7898 \times 10^{-1}}$ (Move the decimal point one place to the right (i.e., $\times 10^1$) and on the right-hand side of the equation, do the opposite of $\times 10^1$, which is $\times 10^{-1}$.)

$\times 10^1$

(c) $830000. = 8.3 \times 10^5$ (Move the decimal point five places to the left (i.e., $\times 10^{-5}$) and on the right-hand side of the equation, do the opposite of $\times 10^{-5}$, which is $\times 10^5$.)

$\times 10^{-5}$

Lesson 26: Scientific Notation

1 1 6

(d) .004039 $= 4.039 \times 10^{-3}$ (Move the decimal point three places to the right (i.e., $\times 10^3$), and on the
right-hand side of the equation, do the opposite of $\times 10^3$, which is $\times 10^{-3}$.)

$\times 10^3$

Note above, for example, that: in (d) .004039

$$= .004039 \times 10^3 \times 10^{-3} \qquad \text{(Note: } 10^3 \times 10^{-3} = 10^0 = 1)$$
$$= (.004039 \times 10^3) \times 10^{-3}$$
$$= 4.039 \times 10^{-3}$$

Lesson 26 Exercises

A. Write in scientific notation: (*a*) 7856.3; (*b*) 890400; (*c*) .000562; (*d*) 45.35; (*e*) 0.0046

Answers: (*a*) 7.8563×10^3; (*b*) 8.904×10^5; (*c*) 5.62×10^{-4}; (*d*) 4.535×10; (*e*) 4.6×10^{-3}

B. Simplify and leave answer in scientific notation:

1. $[9.3 \times 10^5] \times [4.0 \times 10^{-8}]$ **2.** $\dfrac{[8.1 \times 10^4] \times [5.1 \times 10^{-8}]}{[1.7 \times 10^{-7}] \times [9.0 \times 10^2]}$

Answers: 1. 3.72×10^{-2}; **2.** 2.7×10.

Lesson 27
Measurements
(see also p. 165)
American and Metric Systems
Measurement of Length, Mass and Volume

When one goes shopping to buy fabric, the length of the fabric has to be determined. When one goes to buy meat, the weight (mass) of the meat has to be determined, and when one goes to buy gasoline or orange juice, the volume of the gasoline or orange juice has to be determined. For the fabric, we measure its length, for the meat we measure its mass (weight), and for the gasoline or orange juice, we measure its volume. In everyday life as well as in science, we therefore measure length, mass (weight) and volume.

There are at present two main systems of measurement, namely the American system (or the British system) and the metric system. At present, most countries in the world use the metric system of measurement. However, in the United States, most measurements are made using the American system, except in science.

In measurements, we may have to convert from one unit to another unit within the American system or metric system; or we may have to convert from a unit in the American system to a unit in the metric system. We therefore need relationships between units **within** each system, which we shall call **intraconversion factors** and relationships between units **not in** the same system, which we shall call **interconversion factors**. On pages 119-120, we present two charts for some of the basic conversion factors for the two systems. Below, two conversion examples are presented.

Conversion from one unit to another unit

Converting from one unit to another unit is a direct proportion problem, and therefore all the techniques for solving proportion problems we have learned so far are applicable. In particular, the Units Label method (page 65) is highly recommended, especially, for conversions in which we use two or more conversion factors. Below, we apply the Units Label method (Review page 65).

Case 1

Example 1   Convert 96 in. to yd. (Intraconversion: conversion within the American system)

Solution   Plan: This plan is based on what is available in the conversion chart or table.

> (Using a conversion chart is analogous to using a map to go from one location to another)
> The sequence is inches--> feet---> yards.
> We use these relationships: 12 in. = 1 ft;   3 ft = 1 yd (We could use 36 in = 1 yd) (Note: inch = in.)

Step 1: $\dfrac{96 \text{ in.}}{1} \times \dfrac{?}{?}$

Step 2: Multiply the $\dfrac{96 \text{ in.}}{1}$ by a fraction formed by using the quantities 12 in. and 1 ft as the terms of the fraction such that the denominator has the same units as the 96 in.; followed by similar multiplication by a fraction using 3 ft and 1 yd as the terms of the fraction. Then, we obtain

$$\frac{96 \text{ in.}}{1} \times \frac{1 \text{ ft}}{12 \text{ in.}} \times \frac{1 \text{ yd}}{3 \text{ ft}} \qquad \left(\frac{96 \text{ in.}}{1} \times \frac{1 \text{ ft}}{12 \text{ in.}} = \frac{96 \text{ in.}}{1} \times \frac{1 \text{ ft}}{12 \text{ in.}} \times \frac{1 \text{ yd}}{3 \text{ ft}} = \frac{96 \times 1 \times 1 \text{ yd}}{1 \times 12 \times 3} \right)$$

$$= 2\tfrac{2}{3} \text{ yd}$$

$$\therefore \ 96 \text{ in.} = 2\tfrac{2}{3} \text{ yd}$$

Case 2

Example 2: Convert 96 inches to meters (Interconversion: from the American system to the metric system))

Solution Plan: The sequence is inches ----> centimeters --->meters

(The expressions within the parentheses are the steps for the expression on the left.)

$$\frac{96 \text{ in}}{1} \times \frac{2.54 \text{ cm}}{1 \text{ in}} \times \frac{1 \text{ m}}{100 \text{ cm}} \quad \left(\frac{96 \text{ in}}{1} \times \frac{2.54 \text{ cm}}{1 \text{ in}} = \frac{96 \text{ in}}{1} \times \frac{2.54 \text{ cm}}{1 \text{ in}} \times \frac{1 \text{ m}}{100 \text{ cm}} = \frac{96 \times 2.54 \text{ m}}{100} \right)$$

= 2.44 m

∴ 96 in. = 2.44 m (to the nearest hundredth)

Conversion Factors for Measurements

American System (British System) Interconversion (Factors) Metric System

Some "**bridges**" for converting from one system to the other

Length

12 inches (in) = 1 foot (ft.)
3 feet (ft.)   = 1 yard (yd)
5280 feet   = 1 mile (mi)
1760 yards   = 1 mile

1 in. = 2.54 cm
1 yd = 0.9144 m
1 mi = 1.61 km
1 km = 0.62 mi

1 kilometer (km) $= 10^3$ m = 1000 m
1 hectometer (hm) $= 10^2$ m = 100 m
1 dekameter (dam) $= 10^1$ m = 10 m
1 meter (m) $= 10^0$ m = 1m
1 decimeter (dm) $= 10^{-1}$ m = 0.1m
1 centimeter (cm) $= 10^{-2}$ m = 0.01m
1 millimeter (mm) $= 10^{-3}$ m = 0.001m

Some "**bridges**" for converting from one system to the other

Mass

1 lb = 16 oz
1 ton = 2000 lb
1 long ton = 2240 lb

1 kg = 2.2 lb
1 lb = 454 g
1 oz = 28.4 g = 16 drams
1 ton = 0.9072 metric ton

1 kilogram (kg) $= 10^3$ g = 1000 g
1 hectogram (hg) $= 10^2$ g = 100 g
1 dekagram (dag) $= 10^1$ g = 10 g
1 gram (g) $= 10^0$ g = 1 g
1 decigram (dg) $= 10^{-1}$ g = 0.1g
1 centigram (cg)) $= 10^{-2}$ g = 0.01g
1 milligram (mg) $= 10^{-3}$ g = 0.001g

Some "**bridges**" for converting from one system to the other

Volume

16 fluid oz (fl-oz)= 1 pint (pt)
2 pints (pt) = 1 quart (qt)
4 quarts   = 1 gallon (gal)

1 liter (l) = 1.057 qt
1 gal = 3.785 l
1 liter = 2.1 pt
1 pt = .473 l

1 kiloliter (kl) $= 10^3\, l$ = 1000 l
1 hectoliter (hl) $= 10^2\, l$ = 100 l
1 dekaliter (da$l$) $= 10^1\, l$ = 10 l
1 liter (l) $= 10^0\, l$ = 1 l
1 deciliter (dl) $= 10^{-1}\, l$ = 0.1l
1 centiliter (cl) $= 10^{-2}\, l$ = 0.01l
1 milliliter (ml) $= 10^{-3}\, l$ = 0.001 l

Mnemonic device (metric system)

Must remember the following (metric system:)

100 cm   = 1 m
1000 m = 1 km

1000 mg = 1 g
1000 g   = 1 kg

1000 ml = 1 l
1 ml = 1 cc = 1cm^3
1000 cc = 1 l

k – ilo – 10^3
h – ecto – 10^2
d – eka – 10^1
d – eci – 10^{-1}
c – enti – 10^{-2}
m – illi– 10^{-3}

Say the following aloud:
Step 1: First go down vertically as kei-eitch-dii-dii-see-em, then Step 2
Step 2: Kilo-hecto-deka-deci-centi-milli, and then note how the powers
     decrease vertically downwards.
Examples: 1 Kilometer = 10^3 meter; 1 milligram = 10^{-3} gram;
1 centimeter = 10^{-2} meter = $\frac{1}{100}$ meter --->100 centimeters = 1 meter.

Mixed Conversion Factors

Units of length

1 cm = .3937 in = .0328ft = .01094 yd
1 m = 100 cm = 39.3701 in = 3.2808 ft.= 1.0936 yd

1 in. = 2.54 cm = .0833 ft = .0254 m
1 ft = 12 in. = 30.48 cm = .3048 m = .3333 yd
1 mile = 1760 yd = 5280 ft = 1.6093 km
1 yd = 3 ft = 36 in.= .9144 m = 91.44 cm
1 km = .62137 mile = 1000 m = 100,000 cm

Units of mass

1 kg = 1000 g = 2.2046 lb =35.274 oz
1 lb (avdp) = 453.592 g = 16 oz

1 metric ton = 1000 kg = 10^6 g = 1.1023 ton
1 ton = 907.1847 kg = 2000 lb = .9072 metric ton
1 gm = .03527 oz = 1000 mg =.0022046 lb
1 oz = 28.3495 g = .0625 lb = 16 drams
1 long ton (British) = 2240 lb

Units of volume

1 liter = 1000 cm^3 = 61.0237 $in.^3$= .26417 gal = 1.0567 qt =.03531 ft^3= 2.113 pt

1 gal (U.S.) = 4 qt = 3.7854 liter = 8 pt = 231 $in.^3$ = .13368 ft^3

1 qt = 2 pt = .946353 liter =946.353 cm^3 = 57.75 $in.^3$ = .25 gal =.034201 ft^3

1 cord = 128 ft^3
1 pt = .473 liter =.5 qt

1 ft^3 = 1728 $in.^3$

1 yd^3 = 27 ft^3 = 46656 $in.^3$

Units of area

1 ft^2 (sq. ft) = 144 $in.^2$ (sq. in.)

1 yd^2 (sq. yd) = 9 ft^2 (sq. ft) = 1296 sq. in.

1 $mile^2$ (sq. mile) = 640 acres

1 m^2 = $10^4 cm^2$

1 acre = 4840 yd^2

Symbols for units

cm = centimeter	yd = yard	gal = gallon	g = gram
m = meter	km = kilometer	qt = quart	kg = kilogram
in. = inch	mi = mile	pt = pint	lb = pound
ft = foot		oz = ounce	mg = milligram

Some prefixes (International System)

Prefix	Power
tera	10^{12}
giga	10^9
mega	10^6
kilo	10^3
hecto	10^2
deka	10^1
deci	10^{-1}
centi	10^{-2}
milli	10^{-3}
micro	10^{-6}
nano	10^{-9}
pico	10^{-12}

Lesson 27 Exercises

1. Convert 84 in. to m ;   **2.** Convert 1200 cm to yd;   **3.** Convert 4300 m to hm;
4. Convert 6 kg to oz;   **5.** Convert 1760 yd to km;   **6.** Convert 60 pt to ℓ (liters)

Answers: **1.** 2.13 m;   **2.** 13.12 yd;   **3.** 43 hm;   **4.** 211.64;   **5.** 1.61;   **6.** 28.38 ℓ

Test # 2 - Student's Self-Test (Always, Test yourself before you are tested)

Attempt all questions on clean sheets of paper, **Do not write in the book** Show all necessary work.

---

1. Write as decimals:

(a) $\dfrac{56}{1000}$; (b) $\dfrac{60}{100}$.

2. Add: $53.050 + 73.2 + 9 + .0504 + 26$

---

3. Subtract: $58 - 14.603$

4. What must be added to 2.53 to obtain 8?

---

5. Simplify: (a) 62.8×100; (b) 4.206×10^2.

6. Simplify: (a) $54.06 \div 10^4$; (b) $\dfrac{364.8}{10000}$

---

7. Determine the smallest decimal:

(a) .028, .0098, .12706

8. Write in scientific notation:

(a) 7846.3; (b) .000832;

---

9. In the bar graph below, find the ratio of the 1980 population to the 1982 population.

10. Simplify and leave answer in scientific notation:

$$[9.4 \times 10^5] \times [5.0 \times 10^{-8}]$$

---

11. Change $\dfrac{2}{7}$ to a decimal, rounding off quotient to the nearest tenth.

12. Change each decimal to a fraction in its lowest terms

(a) 0.048; (b) 0.0725

---

13. (a) Find 60% of 35;

(b) Find $6\frac{2}{9}\%$ of 2700.

14. 18% of what number is 72?

---

15. The monthly rent for Maria's apartment is $600. If Maria spends 20% of her monthly income on this rent, what is her monthly income?

16. What rate% of 16 is 64?

17. Last year, a homeowner borrowed money from a bank at interest rate of 10% per year. If the homeowner is to pay the bank an interest of $5,000 after one year, how much money did the homeowner borrow?

18. A new math textbook sells for $30.00. However, a used edition of this book is sold at a 5% discount. What is the selling price of the used edition?

19. Last year, the president of a corporation was earning $180,000 per year. This year, because of financial problems, the annual salary is to be reduced to $150,000. What is the percent decrease in salary?

20. A student's grades for class tests were 85, 79, 0, 83, and 78. What was the student's average class grade ?

21. Simplify: $58 + 5 \times (2 + 4)^2$

22. The ratio of male students to female students at a certain college is 3:5. if there are 2,400 students at this college, how many of the students are females?

23. (a) Find the area of the figure below.

(b) Find the perimeter of the figure.

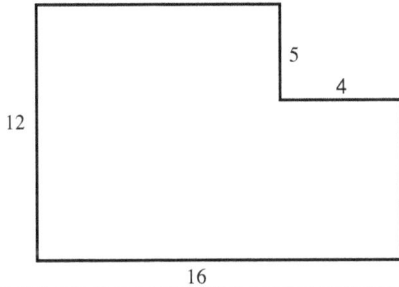

24. (a). Find the area of the figure below.

(b). Find the perimeter of the figure below.

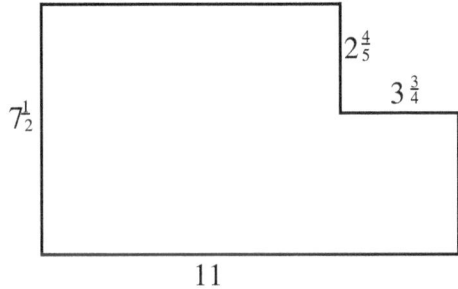

25. Simplify and leave answer in scientific

notation: $\dfrac{[2.1 \times 10^4] \times [2.7 \times 10^{-8}]}{[1.4 \times 10^{-7}] \times [9.0 \times 10^2]}$

Bonus:

James and Mary are to divide $84 in the ratio 1:2. How much does each receive?

Answers: 1 (a) .056 ; (b) .60 ; **2.** 161.3004 ; **3.** 43.397 ; **4.** 5.47 ; **5.** (a) 6280; (b) 420.6;

6. (a) .005406; (b) .03648; **7.** .0098 ; **8.** (a) 7.8463×10^3; (b) 8.32×10^{-4}

9. Ratio 5:6 ; **10.** 4.7×10^{-2} ; **11.** 0.3 ; **12.** (a) $\dfrac{6}{125}$; (b) $\dfrac{29}{400}$; **13.** (a) 21; (b) 168 ;

14. 400 ; **15.** $3,000 ; **16.** 400% ; **17.** $50,000 ; **18.** $28.50 ; **19.** $16\frac{2}{3}\%$; **20.** 65 ;

21. 238 ; **22.** 1,500 ; **23.** (a) 172 sq. units; (b) 56. units.; **24.** (a) 72 sq. units; (b) 37 units ;

25. 4.5×10^0 ; Bonus: James receives $28, and Mary receives $ 56.

Test # 3 - Class Test

Student's Name.....................................

Student's ID.................................

Attempt all questions on the spaces provided. Show all necessary work

---

1. Simplify: $\frac{1}{2} \div \frac{1}{4} \times \frac{4}{5}$

2. Evaluate: $7 + 4 \div 2 \times 3^2$

---

3. Write using digits:
   One hundred seventy-two and nine hundredths.

4. Simplify: $43\frac{1}{4} - 18\frac{5}{8}$

---

5. Simplify: $16\frac{3}{4} + 28\frac{2}{3}$

6. Simplify: $\frac{2}{3} + \frac{5}{6} + \frac{7}{8}$

7. Find (a) the perimeter and
      (b) the area of a rectangle of
      length 8 inches and width 4 inches.

8. Round off 7,893,200 to the nearest thousand.

9. Write 420.05 in words.

10. Change $\frac{6}{7}$ to a decimal rounded off to the nearest hundredth.

11. Add the following:
    68.03;   142; 7.875; and .0832

12. Find the smallest and the largest.:
    .08;   .3; 1.004; .009;   .08001

13. 24 is 30% of what number?

14. Change $4\frac{2}{3}\%$ to a fraction

15. Solve for *x:* $\frac{x}{6} = \frac{5}{2}$

16. A bag which originally sold for $24.50 is now being sold at a 20% discount. What is the present price?

17. Divide: $20.24 \div 4.6$

18. Multiply: .37 by 2.8

19. Maria got grades of 86, 92, 100, 88, and 94 on her tests. What was her average grade.?

20. Simplify: $\left(\dfrac{3}{4} \div \dfrac{1}{2}\right) \div 1\dfrac{1}{5}$

Answers: 1. $1\frac{3}{5}$; **2.** 25 ; **3.** $172\frac{9}{100}$ or 172.09; **4.** $24\frac{5}{8}$; **5.** $45\frac{5}{12}$; **6.** $2\frac{3}{8}$; **7.** (a) 24 in.; (b) 32 sq. in. ; **8.** 7,893,000; **9.** Four hundred twenty and five hundredths ; **10.** 0.86 ; **11.** 217.9882; **12.** Smallest: .009; largest: 1.004; **13.** 80 ; **14.** $\frac{7}{150}$; **15.** $x = 15$; **16.** $19.60 ; **17.** 4.4; **18.** 1.036 ; **19.** 92 ; **20.** $1\frac{1}{4}$.

Test # 4 - Class Test

Student's Name..................................
Student's ID................................

Attempt all questions on the spaces provided. Show all necessary work

---

1. Find x from the proportion: $\frac{x}{5} = \frac{7}{15}$

2. If 80 pounds of a material cost \$1.60, what is the cost of 320 pounds of the same material?

---

3. Express 65% as a fraction in its lowest terms.

4. Change $\frac{7}{25}$ to a percent.

---

5. Find 8% of 2400.

6. Simplify: $7 + 6 \div 3 \times 2$

7. Find $6\frac{2}{9}\%$ of 3600

8. Divide: $324.07 \div 10^2$

9. Change .64 to a fraction in its lowest terms.

10. Change 4.8% to a fraction in its lowest terms

11. 56 is 25% of what number?

12. Multiply: 23.46×10^3

13. Simplify: .07894 × 100

14. Change $\frac{68.29}{1000}$ to a decimal.

15. Show whether or not the following is a
proportion: $\frac{6}{9} = \frac{8}{12}$?

16 Change .35 to a fraction in its lowest terms.

17. Multiply: 6.23 × .47

18. Divide: .56 ÷ .7

19. What rate% of 36 is 108?

20. Subtract 7.146 from 34

Answers: **1.** $2\frac{1}{3}$; **2.** $6.40 ; **3.** $\frac{13}{20}$; **4.** 28% ; **5.** 192 ; **6.** 11 ; **7.** 224 ; **8.** 3.2407 ;

9. $\frac{16}{25}$; **10.** $\frac{6}{125}$; **11.** 224 ; **12.** 23460 ; **13.** 7.894 ; **14.** .06829 ; **15.** Yes ; **16.** $\frac{7}{20}$;

17. 2.9281; **18.** 0.8 ; **19.** 300% ; **20.** 26.854 .

Test # 5 Student's Self-Test (Always, Test yourself before you are tested)

Attempt all questions. Show all necessary work **Word Problems**

1. Maria has $2048 in her savings account and $900 in her checking account. Both accounts are at the same bank. She wants to write a check against her checking account for $1,500.

(a) At least, how much money should Maria transfer from the savings account to the checking account in order to have sufficient funds in the checking account?

(b) After this transfer, what are the balances in the savings and checking accounts?

(c) After this transfer, what percent (approximately) of the total deposit at this bank is in her savings account.

(d) Before the above transfer, what percent (approximately) of the total deposit was in her checking account.

2. 300 ml of a 10% solution of hydrochloric acid are available for a chemistry class experiment. If each student needs 15 ml of this 10% solution for his/her experiment, how many students can perform the experiment.

3. The population of a city is 3,500,000. Of this population, 1,200,000 are males.

(a) How many females live in this city?

(b) What is the ratio of males to females?

(c) Approximately, what percent of the population are females?

4. Maria has $200 to purchase her books. She buys 5 notebooks at $1.20 each, a math textbook at $55, a chemistry textbook at $60 and a biology textbook at $50.00. How much money is left over after these purchases?

5. Janet bought 4 cans of soda for $2, and a pound of sugar for $1. If she paid for these items with a $5.00 bill, how much change does she expect?

6. An engineer's monthly salary a year ago was $3,000. This year, the engineer's salary is $3,150 ` per month.

(a) What is the percent increase in monthly salary.

(b) How much more money will the engineer earn this year than last year?

7. Albert weighs 6 kilograms and Joana weighs $8\frac{1}{4}$ kilograms. By how much does Joana weigh more than Albert?

8. If 4 drams of a drug are given to a person weighing 20 pounds, what will be the weight of a person who is given 12 drams of this drug?

9. How many subway tokens at $1.50 can one buy for $13.50?

10. How many subway tokens at $1.50 each can one buy if one has $10?

11. A rectangular room is 12 feet long, 10 feet wide and 9 feet high.
   (a) Find the area of the floor.
   (b) Find the total area of the walls of the room
   (c) If carpet costs $5.00 per square foot, what is the cost of carpet needed to cover the floor of this room?

12. What is the area of a square of side 4.5 feet?

13. A box contains 90 marbles in two colors, namely, red and black. 20% of these marbles are red
   (a) What is the ratio of the number of red marbles to the number of black marbles?

   (b) How many red marbles must be added so that 70% of the marbles are red?

   (c) How many red marbles must be added so that the ratio of red marbles to black marbles is 2:3?

14. James deposited $2,000 in his savings account for one year at interest rate of $5\frac{1}{4}$% per year
   . What is the balance in this savings account after one year?

Answers: 1. (a) $600; (b) Savings: $1448. Checking: $1500; (c) 49%; (d) 31% ;

 2. 20 students ; **3**. (a) 2,300,000 females ; (b) 12:23; (c) 66%; **4**. $29 ; **5**. $2; **6**. (a) 5%;

(b) $1800 ; **7**. $2\frac{1}{4}$ kilograms ; **8**. 60 pounds ; **9**. 9 tokens ; **10**. 6 tokens ;

11. (a) 120 sq. ft. ; (b) 396 sq. ft.; (c) $600; **12**. 20.25 sq. ft ; **13**. (a) 1:4 ; (b) 150 red marbles
 ; (c) 30 red marbles ; **14**. $2,105.

APPENDIX A
Elementary Algebra
Signed Numbers and Real Number Operations

Lesson 28: **Addition and Subtraction of Signed Numbers**

Lesson 29: **Multiplication and Division of Signed Numbers**

Lesson 30: **Operations Involving Zero, Powers and Roots of Numbers**

Lesson 28
Addition and Subtraction of Signed Numbers

Signed Numbers

A signed number is a number with either a plus sign " +" or a minus sign "-" preceding it (in front of it). If there is no sign in front of a number , we will assume that the number has a plus sign. We call a number with a plus sign a positive number, and we call a number with a minus sign a negative number. For a positive number, we sometimes do not write the plus sign, but for a negative number we **must** always write the minus sign.

The Real Number Line and Real numbers

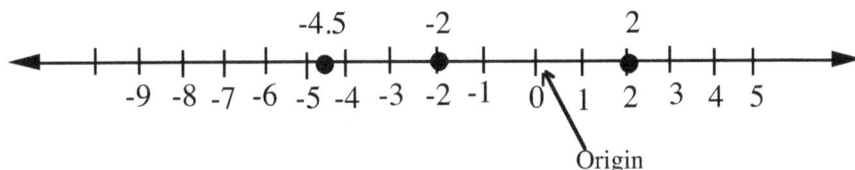

Figure 1

The real number line is a horizontal straight line with equally spaced intervals as in Figure 1 above. We label a point called the origin, 0 (zero). Points to the right of the origin are labeled positive and points to the left of the origin are labeled negative. The numbers increase as one moves from the left to the right on the real number line. Roughly speaking, a real number is a number that can be represented by a point on the real number line. The real numbers consists of the integers, fractions, mixed numbers, decimals, and radicals. In Figure 1, if the real numbers, -4.5, -2, and 2 are of interest, we can represent them by the dots shown. Every point on this line is associated with a real number; and every real number is associated with a point on this line. We can also say that the set of real numbers consists of the signed numbers and zero.

Absolute Value

The absolute value of a signed number may be defined as the number without its sign.

Examples: **1**. The absolute value of -4 , symbolized $|-4| = 4$.

        2. The absolute value of 6, symbolized $|6| = 6$.

        3. The absolute value of $+9$ is 9.

        4. The absolute value of 0 is 0.

Note: The absolute value of a signed number is also its distance from zero on the number line.

Addition of Two Signed Numbers

To add any two signed numbers, first determine which of the following two cases is involved.

Case 1: The two numbers have the same sign
> Step 1: Add the absolute values of the numbers (the unsigned parts)
> Step 2: Write the common sign in front of the sum from Step 1.

Examples   **1.** $(+2) + (+3) = +5$ or 5

2. $(-3) + (-4) = -7$

3. $(-.5) + (-.25) = (-.75)$

Case 2: The two numbers have different signs

Step 1 : Subtract the smaller absolute value from the larger absolute value
> (This subtraction is exactly like the subtraction in arithmetic, where we always subtract a smaller number from a larger number)

Step 2. Write the sign of the number with the larger absolute value in front of the difference (result) from Step 1.

Examples   **1.** $(+5) + (-7) = -2$

2. $(-6) + (+9) = +3$ or 3

3. $(-8) + (+6) = -2$

Note that   **1.** $-8 + 6$ implies $(-8) + (+6) = -2$

2. $-6 + 9$ implies $(-6) + (9) = +3$ or 3

3. $-8 + 9 - 7$ implies that we are to add the numbers, with each number carrying the sign in front of it, and thus
$-8 + 9 - 7 = (-8) + (+9) + (-7) = -6$

Absolute value defined more formally:

The absolute value of a real number x is x if x is a positive number or zero, but it is $-x$ if x is a negative number . (i.e. the negative of a negative number).

Subtraction of Signed Numbers

Procedure:

Change the sign of the number being subtracted, **and add** the changed number to the number from which you are subtracting (i.e., add the negative of the subtrahend to the minuend).

Examples

1. $(+5) - (+7)$ (subtract)   (Here, +5 is the minuend and +7 is the subtrahend)

Step 1: We change +7 to -7

Step 2: We add +5 and -7   (-7 is the changed number; it is the negative of +7)

$(+5) + (-7)$ (add)

$= -2$

2. $(-6) - (-9)$ (subtract)

Step 1: We change -9 to +9

Step 2: We add -6 and +9   (+9 is the changed number)

$(-6) + (+9)$ (add)

$= +3$

3. $-7 - (-10)$ (subtract)

Step 1: We change -10 to + 10

Step 2: We add -7 and + 10   (+10 is the changed number)

$-7 + (+10)$ (add)

$= +3$

4. $8 - (12)$ (subtract)

Step 1: We change 12 to -12

Step 2: We add 8 and -12   (-12 is the changed number)

$8 + (-12)$ (add)

$= -4$

Note the following about the plus "+" and minus "-" signs.

The plus sign "+" may imply either the sign of a positive number, for
example, +2, positive 2; or it may imply the operation of addition, for
example, (+2) + (-8). ——Add

The minus sign "-" may imply the sign of a negative number,
for example,-6, negative 6, or the operation of subtraction, for
example, (-6) - (+7). ——Subtract

Note also that whenever we are to "add" we must use the rules for the addition of signed numbers and
to subtract we must use the subtraction rule(s) for signed numbers.

Subtraction defined more formally:

To subtract a signed number y from another signed number x means add the negative of y to x.

that is, $x - y = x + (-y)$.

Lesson 28 Exercises

A. Add: **1.** (-3) + (-9);   **2.** (+5) + (+2);   **3.** (-7) + (-6)

Answers **1.** -12;   **2.** +7 or 7 ;   **3.** -13

B. Add: **1.** (-8) + (3);   **2.** (5) + (-9);   **3.** (7) + (-4)

Answers   **1.** -5;   **2.** -4 ;   **3.** +3 or 3

C. Simplify: **1.** 5 - 7 ;   **2.** - 4 - 7;   **3.** -8 + 3 - 2;   **4.** -3 + 8;   **5.** -2 + 9 + 3;

Answers **1.** -2 ;   **2.** -11 ;   **3.** -7;   **4.** 5 ;   **5.** 10.

D. Perform the indicated operations:

1. (+7) + (+8) ;   **2.** $-3\frac{2}{5} + 4$;   **3.** (-6) + (-8) ;   **4.** 12 + (-15)

5 . (+9) + (7) ;   **6.** (-3) + (4) ;   **7.** 9 + (8) ;   **8.** -3 + 1

9. $(-2\frac{1}{2}) + (\frac{1}{4})$;   **10.** - 5 + (8) ;   **11.** (4.3) + (-.21);   **12.** 9 - 12

13. (-6) + (-5);   **14.** - 6 + 5 ;   **15.** 8 - 0 ;   **16.** (-3) + (-2)

17. - 4 + 8;   **18.** -1 - 1;   **19.** $-\frac{11}{2} + 3$;   **20.** 5.32 - 7.45

Also find the following:

21. | -2 |=   **22.** | +3 | =   **23.** $|-1\frac{1}{2}| =$ **24.** |-5 | =

Answers: **1.** 15 ; **2.** $\frac{3}{5}$; **3.** -14 ; **4.** -3 ; **5.** 16; **6.** 1; **7.** 17 ; **8.** -2; **9.** $-2\frac{1}{4}$; **10.** 3; **11.** 4.09 ; **12.** -3 ;

13. -11 ; **14.** -1 ; **15.** 8 ; **16.** -5 ; **17.** 4 ; **18.** -2 ; **19.** $-\frac{5}{2}$ or $-2\frac{1}{2}$; **20.** -2.13 ; **21.** 2 ; **22.** 3 ;

23 $1\frac{1}{2}$; **24.** 5

E.
Simplify the following:

1. (-5) - (-7) =   **2.** (-6) - (+7) =   **3.** -6 - (2)   **4.** 6 - (9)

5. -1 - (-1) =   **6.** $-2\frac{1}{2} - (2\frac{3}{4})$ **7.** (8) - (-10)

8. Subtract 6 from 9   **9.** Subtract 9 from 6.   **10.** From 4, subtract 6

11. $\frac{3}{4} - (-\frac{1}{2}))$ **12.** (-4) - (7)   **13.** (-6) - (-8)   **14.** - 8 - (6)

Answers: **1.** 2 ;   **2.** -13;   **3.** -8 ;   **4.** -3 ;   **5.** 0 ;   **6.** $-5\frac{1}{4}$;   **7.** 18 ;   **8.** 3;   **9.** -3;   **10.** -2;   **11.** $1\frac{1}{4}$;
12. -11 ;   **13.** 2 ;   **14.** -14

Lesson 29

Multiplication and Division of Signed Numbers

Multiplication of Two Signed Numbers

Implication of Multiplication   Example: $2 \times 3 = 2(3) = (2)(3) = (2) \cdot (3) = (2)3 = 2 \cdot 3 = 6$

Step 1: Multiply the absolute values.

Step 2: The product is positive (that is, use a plus sign or no sign) if the two numbers have the same sign; but the product is negative (that is, use a minus sign) if the two numbers have different signs.

Examples. The two numbers having the same sign

       1. $(+2)(+3) = +6$ or 6

       2. $(-2)(-3) = +6$ or 6

       3. $-4(-2) = + 8$

The two numbers having different signs

       1. $(-4)(+2) = -8$

       2. $(+5)(-3) = - 15$

       3. $-6(4) = -24$

       4. $(3)(-7) = - 21$

Note the following distinctions:

       1. $-7 - 8 = (-7) + (-8) = -15$ (Addition)

       2. $-7 - (-8) = -7 + (+8) = 1$ (Subtraction)

       3. $-7(-8) = 56$ (Multiplication)

Note also that $(-3)(-2)(-4) = (+6)(-4) = -24$ (The multiplication rules are for **two** numbers at a time.)

Division of Signed Numbers

The rules of signs for division are the same as those for multiplication.

Procedure:

Step 1: Divide the absolute values of the numbers.

Step 2: The quotient is positive (that is, use a plus sign or no sign) if the two numbers have the same sign; but the quotient is negative (that is, use a minus sign) if the two numbers have different signs.

Examples: The two numbers have the same sign

   1. $\frac{+8}{+4} = +2$ or 2;   **2.** $\frac{-8}{-4} = +2$ or 2;   **3.** $\frac{-12}{-9} = +\frac{4}{3})$ or $\frac{4}{3}$

   The two numbers have different signs

1. $\frac{-6}{+2} = -3$;   **2.** $\frac{8}{-2} = -4$;   **3.** $\frac{-13}{3} = -\frac{13}{3}$

Reciprocals: The reciprocal of a real number A is $\dfrac{1}{A}$. Example : The reciprocal of $\frac{2}{3}$ is $\frac{3}{2}$

.Thus, to find the reciprocal of a number, invert the number (or interchange the numerator and the denominator).
 The reciprocal of number is also known as the multiplicative inverse of that number.

The product of a number and its reciprocal is 1. Example $\frac{1}{4} \times \frac{4}{1} = 1$

Application: In $\frac{3}{7} \div \frac{9}{28}$ ($\frac{3}{7}$ is the dividend; $\frac{9}{28}$ is the divisor)

Procedure : Multiply the dividend by the **reciprocal** of the divisor. (Same as invert the divisor and multiply)

$$\frac{3}{7} \div \frac{9}{28} = \frac{3}{7} \times \frac{28}{9} = \frac{4}{3}$$

Lesson 29 Exercises

A. Simplify: **1.** (-5)(-4); **2.** -8(-9) ;   **3.** 6(7) ;   **4.** +5(4); **5.** (-6)(+7); **6.** -4(9)

7. -6(+3);   **8.** 2(-8)

Answers **1.** 20; **2.** 72; **3.** 42; **4.** 20 ; **5.** -42; **6.** -36; **7.** -18; **8.** -16

B. Perform the indicated operations:

1. -8 - 8;   **2.** -8(-8);   **3.** -6 - (8);   **4.** -6 - (-8);   **5.** 9 - 7 + 3 - 8

Answers **1.** -16 ; **2.** 64; **3.** -14   **4.** +2;   **5.** -3

C. Perform the indicated operations:

1. (-4)(-5) =   **2.** (-8)(6) =   **3.** 8(-7) =

4. -8(-7) =   **5.** -1(-5) =   **6.** +6(+4) =

7. 5(6) =   **8.** $4\frac{1}{3}\left(\frac{1}{2}\right)$) =   **9.** (2.3)(.5) =

10. (-4)(+2) =   **11.** (-4)(2) =   **12.** (-3)(-2)(-4) =

Review

13. - 3 - 4 =   **14.** 7 - (-10) =   **15.** -8 - 9 =

16. - 4 + (3) =   **17.** 3 - (-8) =   **18.** 4 - (9) =

19. $5\frac{3}{4} - 8\frac{1}{4}$ =   **20.** (-10) - (-12) =

Answers: **1.** 20 ; **2.** -48 ; **3.** -56 ; **4.** 56 ; **5.** 5 ; **6.** 24 ; **7.** 30; **8.** $2\frac{1}{6}$; **9.** 1.15 ; **10.** -8 ;
11. -8 ; **12.** -24 ; **13** -7 ; **14.** 17; **15.** -17 ; **16.** -1 ; **17.** 11 ; **18.** -5 ; **19.** $-2\frac{1}{2}$; **20.** 2

D. Simplify: **1.** (-6) ÷ (-3);   **2.** $\frac{-12}{-4}$;   **3.** $\frac{-4}{-6}$;   **4.** (-10) ÷ (2);   **5.** $\frac{24}{-6}$;   **6.** $\frac{-15}{3}$

Answers   **1.** 2 ; **2.** 3 ; **3.** $\frac{2}{3}$;   **4.** -5 ; **5.** -4; **6.** -5

E. Carry out the indicated operations:

1. (+12) ÷ (+2);   **2.** (-6) ÷ (-2); **3.** $\frac{14}{-7}$;   **4.** (6) ÷ (-2);   **5.** $-\frac{1}{2} ÷ \frac{1}{4}$;

6. Divide -9 by 2;   **7.** Divide $2\frac{1}{4}$ by $-\frac{1}{4}$;   **8.** Simplify: $\frac{4\frac{1}{5}}{2\frac{1}{5}}$; **9.** Simplify: $\frac{-7.5}{.3}$;

10. Simplify: $\frac{4.28}{-100}$;   **11.** $\frac{+12}{-16}$ **12.** $\frac{-16}{12}$; **13.** Find the reciprocal of $\frac{4}{5}$.

Review: **14.** -4 - 2 =   **15.** 8 - 3 =   **16.** 3 - 8 =

17. -1 - 1 =   **18.** 3 - (-4) =   **19.** 4 - (12) =

Answers: **1.** 6 ; **2.** 3 ; **3.** -2 ; **4.** -3 ; **5.** -2 ; **6.** $-\frac{9}{2}$; **7.** -9; **8.** $1\frac{10}{11}$; **9.** -25 ; **10.** -.0428; **11.** $-\frac{3}{4}$;
12. $-\frac{4}{3}$ or $-1\frac{1}{3}$;   **13.** $\frac{5}{4}$;**14.** -6 ; **15.** 5 ; **16.** -5 ; **17.** -2 ; **18.** 7 ; **19.** -8.

Lesson 30

Operations Involving Zero, Powers and Roots of Numbers

Operations Involving Zero

Addition

Examples 1. $6 + 0 = 6$;   **2.** $0 + 4 = 4$;   **3.** $-7 + 0 = -7$;   **4.** $0 - 5 = -5$;   **5.** $b + 0 = b$

Multiplication

Examples 1. $5(0) = 0$;   **2.** $(7)(9)(0)(14) = 0$

Note: **3.** $0 \times 0 = 0$

Division

Examples 1. $\frac{0}{5} = 0$ (because $5 \times 0 = 0$)

   2. $\frac{0}{-6} = 0$ (because $-6 \times 0 = 0$

But 3. $\frac{5}{0}$ is undefined.(There is **no** number such that $0 \times$ "that number" = 5. Do not divide by 0)

   4. $\frac{0}{0}$ is indeterminate. (Any number will do since $0 \times$ "any number" = 0)

Square root: $\sqrt{0} = 0$

Powers of Signed Numbers

Examples 1. $2^5 = (2)(2)(2)(2)(2) = 32$

2^5 is read as 2 to the fifth power. (or 2 raised to the fifth power)
The exponent "5" indicates how many times the base "2" is being used as a factor.

2. $(-2)^3 = (-2)(-2)(-2)$
     $= -8$
3. $(-3)^2 = (-3)(-3)$
     $= 9$

Zero as an Exponent

Any (nonzero) number raised the to power zero is 1.

Examples 1. $b^0 = 1$ (b $\neq$ 0);   **2.** $4^0 = 1$;   **3.** $(7x^2 dz)^0 = 1$

Note that 0^0 is indeterminate. Example: $0^0 = \frac{0^2}{0^2} = \frac{0}{0}$ which is indeterminate (see above)

Roots of Signed Numbers

Root finding and Power finding are inverse operations (in much the same way as multiplication and division are inverse operations).

Square Root: Symbol " $\sqrt{}$ " or " $\sqrt[2]{}$ "

Finding the square roots of perfect squares by inspection or by factoring

1. We may cheaply define the square root (principal square root) of a nonzero number as one of the two equal **positive** factors of that number. We exclude negative roots.

2. The square root of 0 is 0.

Examples

1. The square root of 9 is symbolized $\sqrt{9}$, and

$$\sqrt{9} = 3$$
$$\text{because } 3^2 = 9$$

Note that , according to the above definition,
$$\sqrt{9} = \sqrt{(3)(3)} = 3$$

2. The square root of 64 is written $\sqrt{64}$, and

$\sqrt{64} = 8$, because $8^2 = 64$.

Also, $\sqrt{64} = \sqrt{(8)(8)} = 8$

3. Note: $\sqrt{0} = 0$, because $0^2 = 0$

If the given number is not a perfect square, we will use tables or a calculator to find the square root.

The **Cube Root** : Symbol" $\sqrt[3]{}$ "

From the above square root definition, we can similarly define the cube root of a number as one of the three equal factors of that number. (In this definition, we do **not** specify **positive root,** since the cube root may be positive or negative, depending on whether the given number is positive or negative.)

Examples

1. $\sqrt[3]{8} = 2$; because $2^3 = 8$.   Note that $\sqrt[3]{8} = \sqrt[3]{(2)(2)(2)} = 2$

2. $\sqrt[3]{-8} = -2$; because $(-2)^3 = -8$.   Note that $\sqrt[3]{(-2)(-2)(-2)} = -2$

More formal definition of square root

1. $\sqrt{A} = r$ if $r^2 = A$, where A, and r are both real and positive.

2. The square root of zero is zero (i.e. $\sqrt{0} = 0$,).

Lesson 30 Exercises

A.. Simplify: **1.** $9 + 0$;   **2.** $3 - 0$;   **3.** $-6 - 0$;   **4.** $(6)(0)$;   **5.** $(0)(-7)$;

6. $\frac{0}{4}$;   **7.** $\frac{0}{-5}$;   **8.** $\frac{9}{0}$;   **9.** $0(0)$;   **10.** $(78)(5)(0)(32)$;   **11.** $\frac{0}{+6}$;   **12.** $0(-8)$

Answers **1.** 9; **2.** 3 ; **3.** -6 ; **4.** 0; **5.** 0 ; **6.** 0; **7.** 0 ; **8.** undefined; **9.** 0 ; **10.** 0 ; **11.** 0 ; **12.** 0

Appendix A Lesson 30: Operations Involving Zero, Powers and Roots of Numbers

B. Evaluate the following:

1. 3^3 2. 2^4 3. 8^2 4. $(-5)^3$

5. $(-1)^3$ 6. $(-1)^4$ 7. $(-1)^5$ 8. $(-\frac{1}{2})^3$

9. 4^0 10. 0^0 11. 0^3 12. $(\frac{3}{2})^2$

13. $(-5)^0$ 14. $(-2)^3$ 15. $(-2)^4$

Review:

16. $-6-6$;   17. $-7(-5)$;   18. $9-(-8)$;   19. $9-8$ 20. $-9-8$

21. $8-9$;   22. $\frac{-16}{4}$ 23. $\frac{0}{-3}$;   24. $\frac{6}{0}$;   25. $(.6)(-.7)$

Answers: **1.** 27; **2.** 16 ; **3.** 64 ; **4.** -125 ; **5.** -1 ; **6.** 1 ; **7.** -1 ; **8** $-\frac{1}{8}$; **9.** 1 ; **10.** indeterminate; **11.** 0 ; **12.** $\frac{9}{4}$;
13. 1 ; **14.** -8 ; **15.** 16; **16.** -12; **17.** 35 ; **18.** 17 ; **19.** 1 ; **20.** -17 ; **21.** -1 ; **22.** -4 ; **23.** 0 ;
24. undefined ; **25.** -.42

C. Evaluate the following : **1.** 4^3;   **2.** $(-3)^4$;   **3.** $\sqrt{64}$;   **4.** $(-5)^3$;   **5.** $\sqrt[3]{-64}$.

Answers **1.** 64 ; **2.** 81 ; **3.** 8 ; **4.** -125 ; **5.** -4

D.

Simplify the following:

1. $\sqrt{25}$ 2. $\sqrt{81}$ 3. $\sqrt{36}$ 4. $\sqrt{256}$

5. $\sqrt{1024}$ 6. $\sqrt[3]{64}$ 7. $\sqrt[4]{81}$ 8. $\sqrt{121}$

9. $\sqrt{\frac{25}{49}}$ 10. $\sqrt[3]{-8}$

Review

11. $-2-2$ 12. $-3(-2)$ 13. $-4-(-7)$ 14. $-2 \div 6$

15. $3(0)$ 16. 3^4 17. $\sqrt{100}$ 18. $6-(-14)$

Answers: **1.** 5 ; **2.** 9 ; **3.** 6 ; **4.** 16 ; **5.** 32 ; **6.** 4 ; **7.** 3 ; **8.** 11 ; **9.** $\frac{5}{7}$; **10.** -2; **11** -4;
12. 6 ; **13.** 3 ; **14.** $-\frac{1}{3}$; **15.** 0 ; **16.** 81 ; **17.** 10 ; **18.** 20.

Test # 1 - Student's Self-Test (Always, Test yourself before you are tested)

Attempt all questions on clean sheets of paper, **Do not write in the book** Show all necessary work.

---

1. (a) $(-4) + (-7) =$

(b) $(-6) + (2) =$

2. $-6\frac{2}{5} + 4 =$

---

3. (a) $(4.3) + (-.27) =$
(b) $-6 - 8 =$

4. (a) $(-6) - (-7) =$

(b) $5 - 20 =$

---

5. (a) Subtract 4 from 7.

(b) $-5 - 12 =$

6. (a) $-4 - (7) =$

(b) $-6(-6) =$

---

7. (a) $(-7)(-3) =$
(b) $(-2)(+6) =$

8. (a) $\left|-4\frac{1}{2}\right| =$
(b) $-7 - (-3) =$

---

9. (a) $8 - 5 + 7 - 9 =$

(b) $(-20) \div (5) =$

10. (a) $4\frac{1}{3}\left(\frac{1}{2}\right) =$

(b) $\frac{-14}{21} =$

---

11. (a) $6\frac{3}{4} - 8\frac{1}{4} =$
(b) $-5(-7) =$

12. (a) $\frac{0}{-7} =$

(b) $-1 - 1 - 1 =$

---

13. (a) $\frac{8}{0} =$
(b) $(-5) + (-4) =$

14. (a) $8^0 =$

(b) $-3(-5) =$

15. (a) $\dfrac{0}{-4} =$

(b) $-1 - 5 =$

16. (a) $\sqrt{49} =$

(b) $- 4 - (2) =$

17. (a) $\sqrt{\dfrac{16}{25}} =$

(b) $4(0) =$

18. (a) $\sqrt{512} =$

(b) $7 - 3 - 6 =$

19. (a) $- 3 - 1 + 9 - 8 =$

(b) $\dfrac{-8}{12} =$

20. (a) $8 - 8 =$

(b) $\dfrac{40}{-16} =$

21. (a) $(-12) \div (4) =$

(b) $(- 7) + (-8) =$

22. (a) $- 1 - 5 + 6 - 8 =$

(b) $-9 - 2 =$

23. (a) $8 - (-5) =$
(b) $(-6) + (-9) =$

24. (a) $-2(-6)(-2) =$

(b) $-2(-6)(-2)(-3) =$

25. (a) $- 6 - (-8) =$
(b) $(+3) + (-7) =$

Bonus: (a) $8 - 7 - 5 =$

(b) $\dfrac{0}{0} =$

Answers: **1.** (a) -11; (b) -4 ; **2.** $-2\frac{2}{5}$; **3.** (a) 4.03; (b) -14 ; **4.** (a) 1; (b) -15 ; **5.** (a) 3; (b) -17; **6.** (a) -11; (b) 36 ; **7.** (a) 21; (b) -12 ; **8.** (a) $4\frac{1}{2}$; (b) -4 ; **9.** (a) 1; (b) -4 ; **10.** (a) $2\frac{1}{6}$; (b) $-\frac{2}{3}$; **11.** (a) $-1\frac{1}{2}$; (b) 35; **12.** (a) 0; (b) -3 ; **13.** (a) undefined; (b) -9 ; **14.** (a) 1; (b) 15 ; **15.** (a) 0; (b) -6 ; **16.** (a) 7; (b) -6 ; **17.** (a) $\frac{4}{5}$; (b) 0; **18.** (a) $16\sqrt{2}$; (b) -2; **19** (a) -3; (b) $-\frac{2}{3}$; **20.** (a) 0; (b) $-\frac{5}{2}$; **21.** (a) -3; (b) -15; **22.** (a) -8; (b) -11; **23.** (a) 13; (b) -15 ; **24** (a) -24; (b) 72 ; **25.** (a) 2; (b) -4; **Bonus:** (a) -4; (b) indeterminate.

APPENDIX B

Order of Operations and Evaluation of Algebraic Expressions

Lesson 31: **Order of Operations in Performing Algebraic operations**

Lesson 32: **Evaluation of Algebraic Expressions Containing Variables**

Lesson 31
Order of Operations in Performing Algebraic Operations

We repeat the order of operations covered previously in Chapter 2; and in the examples covered below, we include evaluations involving negative numbers.

Perform the operations according to the following order
1. Grouping Symbols: If there are grouping symbols, evaluate within the grouping symbols first;
2. Then, evaluate the powers and the roots in any order ; followed by
3. Division and Multiplication from left to right (or invert the divisors and multiply in any order); and then
4. Addition and Subtraction from left to right.

Note: If grouping symbols occur within other grouping symbols, evaluate within the innermost grouping symbols first, followed by the next innermost symbols and so on.

Example 1
$$5 + 2(4 - 7) - 9$$
$$= 5 + 2(-3) - 9 \qquad \text{Evaluating within the parentheses first}$$
$$= 5 - 6 - 9 \qquad \text{Multiplying next}$$
$$= -1 - 9 \qquad \text{Adding from left to right}$$
$$= -10$$

Example 2
$$2^3 - 9(4 + 3) + 8 \qquad \text{Evaluating within the parentheses first}$$
$$= 2^3 - 9(7) + 8$$
$$= 8 - 9(7) + 8 \qquad \textbf{Scrapwork: } 2^3 = (2)(2)(2) = 8$$
$$= 8 - 63 + 8$$
$$= -47$$

Example 3
$$18 \div 6 \times 3 \qquad \text{From left to right, we encounter division first and}$$
$$= 3 \times 3 \qquad \text{therefore, we divide first, according to}$$
$$= 9 \qquad \text{the order of operations.}$$

Note in Example 3 above that, if we had multiplied first, we would **not** obtain the correct answer of 9. In multiplying first, $18 \div 6 \times 3 = 18 \div 18 = 1$ **<-----This is the wrong solution**.

Example 4
$$6 \times 3 \div 18$$
$$= 18 \div 18 \qquad \text{(From left to right, we encounter multiplication}$$
$$= 1 \qquad \text{first, therefore, we multiply first and then divide)}$$
Note however that, if we divide first, we will still get the correct answer.
$$\text{Thus, } 6 \times 3 \div 18 = \frac{6}{1} \times \frac{3}{18} = \frac{6}{1} \times \frac{1}{6} = 1$$

We conclude, from Examples 3 and 4, that if we always divide first (whether or not division comes first) , we will obtain the correct answer; but if multiplication does not come first in going from left to right, then we should **not** multiply first. Thus if we always perform division first, we will obtain the correct answer . Note that Example 3 and Example 4 are different problems.

Appendix B Lesson 31: Order of Operations in Performing Algebraic operations

EXTRA. Simplify: $20 \div \frac{1}{2}$ of 10 (Hint: Perform the "of" operation first before dividing.)

Solution $20 \div \frac{1}{2}$ of 10

     $= 20 \div (\frac{1}{2} \times 10)$ (The notion of changing "of" to multiplication and performing the operations from

     $= 20 \div 5 = 4$ left to right will lead to the **wrong result** in this problem: "of" precedes $\times$ and $\div$.)

Example 5 $10 - [7 - 3(5 - 1)]$
       $= 10 - [7 - 3(4)]$
       $= 10 - [7 - 12]$
       $= 10 - [-5]$
       $= 10 + [+5]$
       $= 15$

                          (Add the terms in the numerator, then

Example 6 $\dfrac{-18 - 6}{9 - 3} = \dfrac{-24}{6}$ add the terms in the denominator before dividing)

           $= -4$

Example 7   Evaluate.

          $0 - 4^2$ $= 0 - (4)(4)$
          $= 0 - 16$
          $= -16$

     Thus, $-4^2 = -16$

Example 8   Evaluate. $0 + (-4)^2$

 Solution: $0 + (-4)^2 = 0 + (-4)(-4)$
                   $= 0 + 16$
                   $= 16$

Thus, $(-4)^2 = 16$; but note that $-4^2 = -16$ (From Examples 7 and 8)

Example 9   Evaluate $(-3)^2 + 7(-3)$

     Procedure: Apply the order of operations.
         $(-3)^2 + 7(-3)$

                                Scrapwork: 1. $(-3)^2 = (-3)(-3) = 9$
      $= 9 - 21$ 2. $9 - 21 = (+9) + (-21) = -12$
      $= -12$

Example 10
          $(-2)^3 - 9(-3)$ **Scrapwork:**   1. $(-2)^3 = (-2)(-2)(-2) = -8$

          $= -8 + 27$ 2. $-9(-3) = +27$
          $= 19$

Example 11
      $3\sqrt{16} - 2(-6) \div 3$
    $= 3(4) + 12 \div 3$
    $= 12 + 4$
    $= 16$

Lesson 31 Exercises

A.

Evaluate the following:
1. $7 - 5(3 + 4)$;   **2.** $8 + 3(5 + 1)^2$;   **3.** $6\sqrt{81} - 5(4)(-2)$

4. $-3[9 - (-4 - 5) + 8]$;   **5.** $15 \div 3 \times 5$ **6.** $12 \times 8 \div 4$

7. $\dfrac{4 - 7}{-8 + 6}$;   **8.** $(-6)^2$;   **9.** -6^2 ;   **10.** $5(-2)^3 - 7(-4)$

Answers: **1.** -28 ; **2.** 116; **3.** 94 ; **4.** -78 ; **5.** 25 ; **6.** 24 ; **7.** $\dfrac{3}{2}$; **8.** 36 ; **9.** -36 ; **10.** -12

B.

Simplify the following:

1. $6 + 2(3 - 8)$ **2.** $4^2 - 3(2 + 5) - 9$ **3.** $12 \div 6 \times 2$

4. $6 \times 2 \div 12$ **5.** $12 - 4[3 - (8 - 2) - 5] + 3$ **6.** $4 - 1[5 + 3]^2 - 8$

7. $0 - 3^2$ **8.** $0 - (-3)^2$ **9.** -3^2 **10.** $(-3)^2$

11. $-2(3) - (-4)$ **12.** $5(-4) + (-3)(2)$ **13.** $-7(-6) - 9(-4)$

14. $(-6)2(0)(8)$ **15.** $3 - 4\{-(6)(-2) + 2[4 - 7]\}$ **16** $\dfrac{-6 - 3}{-2 - 3}$

17. $-3(2 - 8) - (-3)$ **18.** $\sqrt{4^2 + 3^2}$ **19.** $(\frac{2}{3})^3 + \frac{1}{9}$

Review

20. $-8^2 + 3(0)$ **21.** $(-4)^2 - (-3)^2$ **22.** $\dfrac{(-8) - (-2)}{-2}$

Answers: **1.** -4 ; **2.** -14 ; **3.** 4 ; **4.** 1 ; **5.** 47 ; **6.** -68 ; **7.** -9 ; **8.** -9 ; **9.** -9 ; **10.** 9 ; **11.** -2 ;
12. -26 ; **13.** 78 ; **14.** 0; **15.** -21 ; **16.** $\dfrac{9}{5}$ or $1\frac{4}{5}$; **17.** 21 ; **18.** 5 ; **19.** $\dfrac{11}{27}$; **20.** -64 ; **21.** 7 ; **22.** 3.

C.

1. At 7 AM. today, the temperature was 10 degrees below zero. By 3 PM, today, the temperature had increased by 12 degrees. What was the temperature at 3 PM.

2. If an original temperature of 30 degrees above zero decreases by 50 degrees, what is the new temperature?

3. Maria's score on a math test was 4 points above average, while John's score was 3 points below average. What is the difference between Maria's score and John's score?

Answers: **1.** The temperature is 2 degrees above zero;   **2.** The temperature is 20 degrees below zero.
3. The difference is 7 points.

Lesson 32

Evaluation of Algebraic Expressions Containing Variables and Numbers, Given or Knowing the Numerical Values of the Variables

Procedure:
Step 1: Replace each letter by its given numerical value in the expression, (enclosing each value in parentheses during the replacement).
Step 2: Evaluate the resulting expressions using the order of operations learned previously (see p.148).

Example 1   Given that $x = -3$, evaluate $x^2 + 7x + 2$

       Solution: $(-3)^2 + 7(-3) + 2$
                   $= 9 - 21 + 2$ Scrapwork: $(-3)^2 = (-3)(-3)$
                   $= -12 + 2$ $= 9$
                   $= -10$

Example 2   Find the value of $5 - 4[3x - 1]$, if $x = 2$.

Solution   Substituting 2 for x, we obtain
$$5 - 4[3(2) - 1]$$
$$= 5 - 4[6 - 1]$$
$$= 5 - 4[5]$$
$$= 5 - 20$$
$$= -15$$

Example 3   (a) Evaluate: $200\,(2^n)$ when $n = 3$

  Procedure: Substituting n = 3 in $200(2^n)$ and applying the order of operations, then we obtain:

                                    Scrapwork
           $200\,(2^3)$ $2^3 = (2)(2)(2) = 8$
      $= 200\,(8)$
      $= 1600$

(b) Evaluate: $1000\,(4)^n$ if $n = 2$

  Procedure: Substitute $n = 2$ in $1000\,(4)^n$

                                   Scrapwork
            $1000\,(4)^2$ $(4)^2 = (4)(4) = 16$
    $= 1000\,(16)$
    $= 16000$

Lesson 32 Exercises

A. 1. Find the value of $b^2 - 6b + 7$ when $b = -3$

2. Given that $A = \frac{1}{2} bh$, find the value of A if $b = 6, h = 4$

3. Find the value of F if $C = -60$, given that $F = \frac{9}{5}C + 32$

4. Given that $A = L \times W$, find the value of A when $L = 8, W = 6$

Answers   **1.** 34 ;   **2.** 12 ;   **3.** -76;   **4.**   48

B. Evaluate the following if $a = 2, \quad b = 1, \quad c = 3, \quad x = -5, \quad y = -1$

1. $4x + 5y$ **2.** $a^2 - b^2$ **3.** $abcxy$

4. $2ab - xy$ **5.** $-ax - by + 10$ **6.** $8 - a^2 + bc$

7. $(-5)^2 - by$ **8.** $\dfrac{3a - 2}{ax + 2}$

9. Find the value of F if $C = -20$, given that $F = \frac{9}{5}C + 32$

10. Given that $A = \frac{1}{2}bh$, find the value of A when $b = 10$ and $h = 4$.

Answers: 1. -25 ; **2.** 3 ; **3.** 30 ; **4.** -1 ; **5.** 21 ; **6.** 7 ; **7.** 26 ; **8.** $-\frac{1}{2}$; **9.** -4 ; **10.** 20

Test # 2 - Student's Self-Test (**Always**, **Test yourself before you are tested**)   153

Attempt all questions on clean sheets of paper, **Do not write in the book** Show all necessary work.

---

1. (a) $(-2)^4 =$

   (b) $(-3)^3 =$

2. (a) $\sqrt{81}$

   (b) $\sqrt[3]{-8}$

---

3. Evaluate (a) $7 - 2(3 + 4)$

   (b) $16 \times 4 \div 4$

4. (a) $5\sqrt{49} - 3(4)(-2)$

   (b) $-3[8 - (-4 - 3) + 8]$

---

5. $\dfrac{5 - 3}{-10 + 6}$

6. (a) $5 + 3(4 + 1)^2$

   (b) $2(-3)^3 - 7(-5)$

---

7. (a) $(-5)^2 =$

   (b) $-5^2 =$

8. (a) $6 \times 12 \div 2$

---

9. (a) $2 - 4\{ - (4)(-2) + 2[4 - 7]\}$

10. $\dfrac{-6 - 3}{-2 - 3}$

---

11. (a) $0 - 4^2$

   (b) $0 - (-4)^2$

   (c) $0 + (-4)^2$

12. (a) $\sqrt{4^2 + 3^2}$

   (b) $\left(\dfrac{2}{3}\right)^3 + \dfrac{1}{9}$

---

13. (a) $(-4)2(0)(9) =$

   (b) $-7^2 + 2(0) =$

14. (a) $-3(2 - 7) - (-3)$

   (b) $(-5)^2 - (-3)^2$

---

15. (a) $6 - 2[4 + 3]^2 - 8$

16. (a) Find the value of $b^2 - 6b + 7$ when $b = -2$

17. Given that $A = \frac{1}{2}bh$ find the value of A if $b = 10, h = 3$

18. Find the value of F if $C = -20,$ given that $F = \frac{9}{5}C + 32$

19. Given that $A = L \times W,$ find the value of A when $L = 7, W = 4.$

20. $\frac{(-4) - (-1)}{-1}$

Questions 21-25

Evaluate if $a = -2, b = 1, c = 3, x = -5,\ y = -1$

21. $6x + 5y$

22. $abcxy$

23. $\frac{4a - 2}{ax + 2}$

24. $a^2 - b^2$

25   Find the value of F if $C = -40,$ given that $F = \frac{9}{5}C + 32$

.

Bonus:

Answers: **1.** (a) 16; (b) -27 ; **2.** (a) 9; (b) -2 ; **3.** (a) -7; (b) 16 ; **4.** (a) 59; (b) -69 ; **5.** $-\frac{1}{2}$; **6.** (a) 80; (b) -19 ; **7.** (a) 25; (b) -25 ; **8.** 36 ; **9.** -6 ; **10.** $\frac{9}{5}$; **11.** (a) -16; (b) -16; (c) 16 ; **12.** (a) 5 ; (b) $\frac{11}{27}$; **13.** (a) 0; (b) -49 ; **14.** (a) 18; (b) 16 ; **15.** -100 ; **16.** 23 ; **17.** 15 ; **18.** -4 ; **19.** 28 ; **20.** 3; **21.** -35 ; **22.** -30 ; **23.** $-\frac{5}{6}$; **24.** 3 ; **25.** -40.

APPENDIX C

First Degree Equations Containing One Variable
(Simple Linear Equations)

Lesson 33: **Axioms for Solving Equations**
Lesson 34: **Solving First Degree Equations**

Lesson 33
Axioms for Solving Equations

Axioms are general mathematical statements that we accept as true, without any proof, in order to deduce other less obvious statements. The following are very useful in constructing proofs and solving equations.

1. A quantity is equal to itself. (reflexive property of equality, also identity principle)
    $a = a$

2. An equality may be reversed. (symmetric property of equality)
    If $a = b$, then
    $b = a$

3. Quantities equal to the same quantity are equal to each other.(transitive property of equality)
    If $a = b$, and $b = c$, then
    $a = c$.

4. A quantity may be substituted for its equal in any expression or equation. (substitution axiom)

5. A whole equals the sum of all its parts. (partition axiom)

6. If equal quantities are added to equal quantities, the sums are equal (addition axiom)

    If $a = b$,   (If $a = b$, and
    then $a + c = b + c$ **ALSO** $c = d$, then
                                            $a + c = b + d$)

7. If equal quantities are subtracted from equal quantities, the differences are equal.(subtraction axiom)

    If $a = b$,   (If $a = b$, and
    then $a - c = b - c$ **ALSO** $c = d$, then
                                            $a - c = b - d$)

8. If equal quantities are multiplied by equal quantities, the products are equal. (multiplication axiom)

    If $a = b$,   (If $a = b$, and
    then $ac = bc$ **ALSO** $c = d$, then
                                            $ac = bd$.)

9. If equal quantities are divided by equal quantities (not zero), the quotients are equal. (division axiom)

If $a = b$, (If $a = b$, and

then $\dfrac{a}{c} = \dfrac{b}{c}$ **ALSO** $c = d$, then

$$\dfrac{a}{c} = \dfrac{b}{d})$$

10. Like powers of equals are equal . (powers axiom)

Example: If $a = 3$, then

$$a^2 = 3^2 \text{ or } a^2 = 9.$$

11. Like roots of equals are equal. (roots axiom)

Example: if $a^3 = 8$, then

$$\sqrt[3]{a^3} = \sqrt[3]{8}$$
$$a = 2.$$

Lesson 34

Solving First Degree Equations

An equation is a statement of the equality between two expressions.

Example: $2x + 3 = x - 10$

The equality symbol "=" breaks up an equation into two sides or members , namely the left-hand side and the right-hand side of the equation. To solve an equation involving a single variable, say x, means we are to find values of x which satisfy the equation. **A value of x is said to satisfy an equation if this value when substituted in the equation makes both the left-hand side and the right-hand side of the equation equal to each other.** A value of x which satisfies a given equation is said to be a solution or a root of the given equation. **To obtain a value for the variable, x, we will get x by itself on one side of the equation .** We agree that a value of x has been obtained if we have x by itself on one side of the equation and all the other quantities on the other side of the equation do **not** involve x. To get x by itself, we will use inverse operations. Addition and subtraction are inverse operations. Multiplication and division are inverse operations. Inverse operations "undo" each other. For example, to undo multiplication by a number, we will use division by the same the number. We will keep the above discussion in mind when we solve linear equations.

Example 1   Solve for x:

1. $x - 5 = 11$

To get x by itself on the left-hand side of the equation, we will remove the "-5" by adding its opposite to both sides of the equation.(This is the same as adding + 5 to both sides of the equation)

$$\begin{array}{r} x - 5 = 11 \\ +5 \quad +5 \\ \hline x = 16 \end{array}$$

The solution is 16.

Example 2 Solve for x: $x + 8 = -11$

To get x by itself on the left-hand side of the equation, we remove the "+8" by adding -8 to (that is, subtracting 8 from) both sides of the equation.

$$\begin{array}{r} x + 8 = -11 \\ -8 \quad -8 \quad . \\ \hline x = -19 \end{array}$$

(by stressing on **addition** of - 8, the right-hand side of the equation is done without confusion)

Note that to remove (undo) any number, use the inverse (opposite) operation.

Example 3   Solve for x: $5x = 20$

To obtain x by itself, we will undo (remove) the "5". We use division by 5 (since the "5" is multiplying the x)

$$\frac{5x}{5} = \frac{20}{5}$$
$$x = 4$$

The solution is 4.

Example 4 Solve for x:

$$-6x = 24$$
$$\frac{-6x}{-6} = \frac{24}{-6}$$
$$x = -4$$

The solution is -4.

Example 5   Solve for x: $\dfrac{x}{6} = 4$

To remove the "6" which is a divisor, we will multiply both sides of the equation by 6

$$(6)\dfrac{x}{6} = (6)4$$

$$\cancel{(6)}\dfrac{x}{\cancel{6}} = (6)4$$
$$1$$

$x = 24$. The solution is 24.

Example 6 Solve for x: $\dfrac{-x}{5} = 2$ (or $\dfrac{x}{5} = -2 \text{---}{>} x = 5(-2) = -10$)

$$\cancel{(5)}\dfrac{(-x)}{\cancel{5}} = (5)2$$
$$1$$

$-x = 10$ $\leftarrow$ ------------------------

$(-1)(-1x) = (-1)10$ **Note:** $-x = -1x$

$x = -10$ OR   from $\boxed{\text{this step}}$, just change

The solution is -10.   the sign of the left-hand side and
change the sign of the right-hand side.

Example 7   Solve for x
                $2x - 6 = 10$

We will remove the "- 6 " first, and then remove the "2" (i.e., we will always undo multiplication and division last)

$$\begin{array}{r} 2x - 6 = 10 \\ \underline{+6 \quad +6} \\ 2x = 16 \end{array}$$

$$\dfrac{2x}{2} = \dfrac{16}{2}$$

$x = 8$.   The solution is 8.

Example 8   Solve for x: $5x - 6 = x + 2$

In this case, we will collect all the x's on one side by removing either the x-term on the right-hand side or the x-term on the left-hand side.
Let us remove the x-term on the right-hand side.

$$\begin{array}{r} 5x - 6 = x + 2 \\ \underline{-x \qquad -x} \\ 4x - 6 = +2 \\ \underline{+6 \quad +6} \\ 4x = 8 \end{array}$$

$$\dfrac{4x}{4} = \dfrac{8}{4}$$

$x = 2$.   The solution is 2.

Example 9   Solve for x
                $7x + 2 - 3x = 2x + 18 + 3x$

In this case, first, simplify each side of the equation, and then proceed.

Step 1: $7x + 2 - 3x = 2x + 18 + 3x$
                $4x + 2 = 5x + 18$

Step 2: $$\begin{array}{r} \underline{-5x \qquad -5x} \\ -x + 2 = 18 \\ \underline{-2 \quad -2} \\ -x = 16 \end{array}$$

$x = -16$. The solution is -16.

Example 10   Solve for x
$$3(x + 2) = 5(x - 6)$$
First, remove the parentheses on both sides of the equation

$$3(x + 2) = 5(x - 6)$$
$$3x + 6 = 5x - 30$$
$$\underline{-5x \qquad -5x}$$
$$-2x + 6 = -30$$
$$\underline{\quad - 6 \quad - 6}$$
$$- 2x = -36$$
$$\frac{-2x}{-2} = \frac{-36}{-2}$$
$$x = 18$$

The solution is 18.

Checking Solutions of Equations

Example (a)   Determine if - 5 is a solution of the equation
$$4x + 18 - x = 8x + 32 + 2x.$$
- 5 will be a solution of the given equation if this value when substituted in the given equation makes the left-hand side of the equation equal to the right-hand of the equation.

Checking: Replace x by -5 in the equation.

$$\text{Then, } 4(-5) + 18 - (-5) \overset{?}{=} 8(-5) + 32 + 2(-5)$$
$$-20 + 18 + 5 \overset{?}{=} - 40 + 32 - 10$$
$$-2 + 5 \overset{?}{=} - 8 - 10$$
$$3 = -18 \quad \text{False}$$

The left-hand side of the equation and the right-hand side of the equation are **not** equal. Therefore, -5 is **not** a solution of the given equation.

(b) Determine if -2 is a solution of $4x + 18 - x = 8x + 32 + 2x.$

Checking: Replace x by -2 in the equation.

$$\text{Then, } 4(-2) + 18 - (-2) \overset{?}{=} 8(-2) + 32 + 2(-2)$$
$$-8 + 18 + 2 \overset{?}{=} -16 + 32 - 4$$
$$10 + 2 \overset{?}{=} 16 - 4$$
$$12 = 12 \quad \text{True}$$

Since the left-hand side of the equation equals the right-hand side of the equation, -2 is a solution of the given equation.

More Examples on Solutions of Linear Equations

In the following examples, attempt the problems before reading the solutions:

Example 1 Solve for x: $5x - 9 = 6$

Solution

$$5x - 9 = 6$$
$$\underline{+9 \quad +9}$$
$$5x + 0 = 15$$
$$\frac{5x}{5} = \frac{15}{5}$$
$$x = 3$$

Check: $5(3) - 9 \overset{?}{=} 6$

$15 - 9 \overset{?}{=} 6$

$6 = 6$ True

Example 2 Solve for x: $6x - 3 = 2x + 8$

Solution :

$$6x - 3 = 2x + 8$$
$$\underline{-2x \qquad -2x}$$
$$4x - 3 = 0 + 8$$
$$\underline{+3 \qquad + 3}$$
$$4x = 11$$
$$\frac{4x}{4} = \frac{11}{4}$$
$$x = \frac{11}{4}$$

Check: $6(\frac{11}{4}) - 3 \overset{?}{=} 2(\frac{11}{4}) + 8$

$33 - 6 \overset{?}{=} 11 + 16$

$27 = 27$ True

Example 3 Solve for x:

$$\frac{x}{6} = \frac{11}{12}$$

Method 1 To get x by itself on the left-hand side of the equation, we undo the "6" (which is a divisor) by multiplying each side of the equation by 6.

$$(6)\frac{x}{6} = (6)\frac{11}{12}_2$$

$$(\cancel{6})\frac{x}{\cancel{6}}^{1} = {}^{1}(\cancel{6})\frac{11}{\cancel{12}}_2$$

$$x = \frac{11}{2}$$

Method 2 Multiply each term of the equation by the LCD of the fractions. The LCD is 12.

$$(12)\frac{x}{6} = (12)\frac{11}{12}$$

$$2\,(\cancel{12})\frac{x}{\cancel{6}}_1 = (\cancel{12})^{1}\frac{11}{\cancel{12}}_1$$

$$2x = 11$$

$$\frac{2x}{2} = \frac{11}{2}$$

$$x = \frac{11}{2}$$

Method 3 **Cross-multiplication**

$$\frac{x}{6} = \frac{11}{12}$$

$$12x = 6(11)$$

$$\frac{^1\cancel{12}x}{\cancel{12}_1} = \frac{^1\cancel{6}(11)}{\cancel{12}_2}$$

$$x = \frac{11}{2}$$

Note: We can cross-multiply if there is only a single fraction on the left-hand side of the equation, and only a single fraction on the right-hand side of the equation. For instance, in the following example, we cannot cross-multiply immediately, but rather, we will have to combine the right-hand side into a single fraction before cross-multiplying.

Example 4 Solve for x: $\frac{x}{4} = \frac{x}{20} + \frac{2}{5}$

Method 1 The LCD method. Multiply each term of the equation by the LCD of the fractions. The LCD is 20.

$$(20)\frac{x}{4} = (20)\frac{x}{20} + (20)\frac{2}{5}$$

$$^5\cancel{(20)}\frac{x}{\cancel{4}_1} = ^1\cancel{(20)}\frac{x}{\cancel{20}_1} + ^4\cancel{(20)}\frac{2}{\cancel{5}_1}$$

$$5x = x + 8$$

$$\underline{-x \quad -x \qquad} .$$

$$4x = 8$$

$$\frac{4x}{4} = \frac{8}{4}$$

$$x = 2$$

The solution is 2.

Method 2 By cross-multiplication

To apply cross-multiplication to the above problem, first, combine the two terms on the right-hand side into a single fraction. (We have a single fraction if we have a single denominator.)

continuing, $\frac{x}{4} = \frac{x}{20} + \frac{2}{5}$

$$\frac{x}{4} = \frac{x}{20} + \frac{2(4)}{5(4)}$$

$$\frac{x}{4} = \frac{x}{20} + \frac{8}{20}$$

$$\frac{x}{4} = \frac{(x+8)}{20}$$

Now, we cross-multiply:

$$20x = 4(x+8) \;\text{<------ The parentheses are important}$$

$$20x = 4x + 32$$

$$\underline{-\,4x \;-\,4x \qquad}.$$

$$16x = 0 + 32$$

$$\frac{16x}{16} = \frac{32}{16}$$

$$x = 2$$

Method 3   You may undo one denominator at a time.

$$\frac{x}{4} = \frac{x}{20} + \frac{2}{5}$$

Step 1:   Undo the "4" : $$\frac{1\,\cancel{(4)}x}{\cancel{4}_1} = \frac{1\,\cancel{(4)}x}{\cancel{20}_5} + \frac{(4)2}{5}$$ (Multiply each term by 4)

$$x = \frac{x}{5} + \frac{8}{5}$$

Step 2:   To undo the "5" multiply by 5:

$$(5)x = \frac{1\,\cancel{(5)}x}{\cancel{5}_1} + \frac{\cancel{(5)}8}{\cancel{5}_1}$$

$$5x = x + 8$$
$$4x = 8$$
$$x = 2$$

Note: In method 3, multiplying the equation first by 4 and then by 5 is equivalent to multiplying the equation by 20 (as was done in Method 1).

Lesson 34 Exercises

A

Solve for x: **1.** $x - 7 = 12$;   **2.** $x + 6 = 9$; **3.** $4x = 20$;   **4.** $\frac{x}{6} = 8$; **5.** $4x + 3 = 35$

6. $7x - 6 = 2x + 24$;   **7.** $5x - 7 + 3x + 9 = 6x + 5 + 8x$;   **8.** $\frac{x}{6} + 4 = 9$;

9. $3(2x - 5) + 4 = -4(x + 7)$

Answers: **1.** $x = 19$; **2.** $x = 3$; **3.** $x = 5$; **4.** $x = 48$; **5.** $x = 8$; **6.** $x = 6$; **7.** $x = -\frac{1}{2}$

8. $x = 30$; **9.** $x = -\frac{17}{10}$

B   Solve and check:

1. $x + 4 = 6$;   **2.** $x + 5 = -7$ **3.** $x - 3 = 9$

4. $x - 6 = -11$ **5.** $x - \frac{1}{3} = \frac{1}{4}$ **6.** $\frac{x}{2} = 8$

7. $\frac{x}{-2} = 7$ **8.** $\frac{-x}{4} = 9$ **9.** $3x = 18$

10. $-4x = 20$ **11.** $6x = 22$ **12.** $7x = -28$

13. $\frac{x}{4} - 6 = 14$ **14.** $\frac{-x}{3} + 5 = 7$ **15.** $2x + 4 = 9$

16. $-3x - 5 = 10$ **17.** $5x - 4 + x = 4x + 6 - 3x$

18. $2x - 5 + 3x + 2 = 7x - 5 - 3x$ **19.** $2(x - 4) + 3 = 15$;   **20.** $-3(x + 5) + 1 = 5(x + 4) - 6$

Answers: **1.** 2 ; **2.** -12 ; **3.** 12 ; **4.** -5 ; **5.** $\frac{7}{12}$; **6.** 16 ; **7.** -14 ; **8.** -36 ; **9.** 6 ; **10.** -5 ; **11.** $\frac{11}{3}$;

12. -4 ; **13.** 80 ; **14.** -6 ; **15.** $\frac{5}{2}$; **16.** -5 ; **17.** 2 ; **18.** -2 ; **19.** 10 ; **20.** $-\frac{7}{2}$

C

1. Is 2 a solution of $3x + 1 = 7$?   **2**. Is -3 a solution of $4x - 1 = 15$?

3. Is 2 a solution of $5x + 2 = x - 4$?

Answers: **1**. Yes ; **2**. No ; **3**. No.

Solve for x: **1.** $\dfrac{x}{6} + \dfrac{x}{3} = \dfrac{3}{4}$ **2.** $\dfrac{x}{8} = \dfrac{x}{2} - 6$;   **3.** $\dfrac{x}{2} + 1 = \dfrac{x}{3}$;   **4.** $\dfrac{x-2}{4} + \dfrac{2}{3} = \dfrac{1}{2}$

Answers : **1.** $x = \dfrac{3}{2}$; **2.** $x = 16$; **3.** $x = -6$; **4.** $x = \dfrac{4}{3}$

E

Solve for x: **1.** $\dfrac{x+3}{4} + \dfrac{x-2}{8} = 2$ **2.** $\dfrac{x-2}{3} + 1 = \dfrac{x}{5}$ **3.** $\dfrac{x}{4} + \dfrac{1}{2} = \dfrac{x}{5} + 1$

     4. $\dfrac{x}{3} = \dfrac{x}{6} + \dfrac{3}{5}$ **5.** $\dfrac{x}{2} + \dfrac{2}{3} = \dfrac{x}{4}$

Answers: **1.** 4 ; **2.** $-\dfrac{5}{2}$; **3.** 10 ; **4.** $\dfrac{18}{5}$; **5.** $-\dfrac{8}{3}$

APPENDIX D
About Measurements
Standard Unit, Error, Rounding-off Numbers, Significant Digits, Scientific Notation

To determine the size of a physical quantity, we compare its size with a standard quantity called a unit. Example: To determine the length of the cover of a book in inches or in meters, we can use a ruler with its scale in inches or in meters.

A measurement is the ratio of the magnitude of a physical quantity to that of a standard unit

Standard unit

A standard unit is a measure with which other quantities are compared. A standard unit of measure is defined by a legal authority (such as the US Bureau of Weights and Measures) or by a conference of scientists.

Some universally accepted standards:

1. For mass, the standard (primary standard) is the kilogram (kg).
2. For length, the standard is the meter (m).
3. For time, the standard is the second (s)

Some devices for taking measurements

Examples: Rulers (for length), chemical balances (for mass), stop watches (for time), ammeters, (for electric current) voltmeters (for electric voltage) , thermometers (for temperature) and barometers (for pressure).

Experimental Errors (or Uncertainties)

There are two main types of errors, namely, systematic errors, and random errors.

Systematic Errors (constant errors):

These errors are due to faulty measuring devices. Systematic errors make the measurements either too small or too large:

Examples of faulty devices:

1. Instruments with needles off the zero mark; 2. Faulty clocks (stop clock)
3, Corroded weights; 4. Faulty thermometers. Heat leaking equipment

Random errors (accidental errors or indeterminate errors)

Random errors may be due to chance variations of the physical quantity being measured, or chance variations in the measuring device. Random errors may also be due to failure to take into account variables such as temperature fluctuations; and environmental effects. We can reduce random errors by making a large number of measurements and taking the average of the measurements.

Absolute error (or absolute uncertainty)

Absolute error Experimental value − accepted (true value)

Example: In an experiment to determine the acceleration due to gravity, g, the experimental value of g, was 986 cm/s^2. The accepted value of g is 980 cm/s^2. Find the absolute error.

Solution   Experimental value = 986 cm/s^2

Accepted value = 980 cm/s^2

Absolute error = experimental value - accepted value

= (986 - 980) cm/s^2

= 6 cm/s^2

The positive value indicates that the experimental value is greater than the accepted value.

Note: If the experimental value = 964 cm/s^2

The absolute error = (964 - 980) cm/s^2

$$= -16 \text{ cm/} s^2$$

The negative value means that the experimental value is less than the accepted value.

Relative error (or relative uncertainty)

$$\text{Relative error} = \frac{\text{absolute error}}{\text{accepted value}}$$

Example 1: If the absolute error = 6 cm/s^2, and

the accepted value = 980 cm/s^2,

$$\text{Relative error} = \frac{\text{absolute error}}{\text{accepted value}}$$

Then relative error = $\dfrac{6 \ cm / s^2}{980 \ cm / s^2}$

$$= 0.00612$$

Example 2: If the absolute error = -16 cm//s^2 and

the accepted value = 980 cm/s^2

Then relative error = $\dfrac{-16 \ cm / s^2}{980 \ cm / s^2}$

$$= - 0.00612$$
$$= - 0.0163$$

Note: If an accepted value is not known, and we have two or more experimental values,
then the average of the experimental values would be used as the "accepted value" in calculations.

Rounding-Off Numbers

The rules for rounding-off a number may differ slightly depending upon the field.
For instance, in accounting the rule may be slightly different from the rule in chemistry.

1. If the digit or group of digits to be dropped is more than 500...,hen drop that digit or group and add 1 to the last digit retained.

2. If the digit or group of digits to be dropped is less than 500...,then drop that digit or group and leave the last digit retained unchanged.

3. If the digit or group of digits to be dropped is exactly 500..., then drop that portion and add 1 to the last digit retained if this digit is odd but if this digit is even, then this digit remains unchanged.

Rounding off Whole numbers

Procedure:

Step 1: Locate the digit in the round-off place.
(The round-off place is the place to which we want to round-off the number)

Step 2: Drop all digits to the right of the round-off place, and if the digit immediately to the right of the round-off place is more than 5 or is 5 followed by non-zero digits, , add 1 to the round-off place digit (i.e. we round-up); but if the digit immediately to the right of the round-off place is less than 5, the round-off place digit remains unchanged. However, if the digit immediately to the right of the round-off place digit is 5 or 5 followed by zeros, we add 1 to the round-off place digit if it is odd, but if it is even, it remains unchanged.(i.e. we round-down). Also, replace each digit dropped by a zero.

Rounding-off Decimals

The procedure is the same as that for rounding-off whole numbers, except that after the decimal point, we do not replace any digits dropped by zeros.

Procedure:

Step 1: Locate the digit in the round-off place.
(The round-off place is the place to which we want to round-off the number)

Step 2: Drop all digits to the right of the round-off place, and if the digit immediately to the right of the round-off place is more than 5 or is 5 followed by non-zero digits, add 1 to the round-off place digit (i.e., we round-up); but if the digit immediately to the right of the round-off place is less than 5, the round-off place digit remains unchanged. However, if the digit immediately to the right of the round-off place digit is 5 or 5 followed by zeros, we add 1 to the round-off place digit if it is odd, but if it is even, it remains unchanged.(i.e., we round-down).

Rounding-off (Alternatively)

When we round-off a number, we drop some of the digits explicitly or implicitly specified.
We must distinguish between rounding-off to a specified number of decimal places
or significant digits) and the implicit rounding-off which we must determine from the
numbers involved in the calculation.

The rules for rounding-off a number may differ slightly depending upon the field.
 For instance, in accounting the rule may be slightly different from the rule in chemistry.

1. If the digit or group of digits to be dropped is more than 500...then drop that
      digit or group and add 1 to the last digit retained.

2. If the digit or group of digits to be dropped is less than 500...then drop that digit or
      group and leave the last digit retained unchanged.

3. If the digit or group of digits to be dropped is exactly 500..., then drop that portion and
      add 1 to the last digit retained if this digit is odd but if this digit is even, then this digit
      remains unchanged.

Example: The following have been rounded-off to three decimal places.

(1)   .4398. ≈ **.440**

(2)   .43652 ≈ **.437**

(3)   .43637 ≈ **.436**

(4)   .43750 ≈ **.438**

(5)   .43650 ≈ **.436**

(6)   .43946 ≈ **.439**

(7)   .43650001 ≈ **.437**

(8)   .4365001 ≈ **.437**

Example We round off **85376.7463** to the following places, using the simple "5 or greater
 or less than 5 rule"

1. 85376.7463 to the nearest **thousandth** becomes **85376.746** (We do **not** replace the 3 dropped by a zero)

2. 85376.7463 to nearest **hundredth** becomes **85376.75** (We added 1 to the digit in the round-off place)

3. 85376.7463 to the nearest **tenth** becomes **85376.7** (The 7 is unchanged since the 4 dropped is less than 5)

4. 85376.7463 the nearest **unit** becomes **85377.**   (Adding 1 to the 6)

5. 85376.7463 to the nearest **ten** becomes **85380.** (Replacing the 6 dropped by a zero)

6. 85376.7463 to the nearest **hundred** becomes **85400.** (Replacing the digits (6 and 7) dropped by zeros)

7. 85376.7463 to the nearest **whole number** becomes **85377**. (same as to the nearest unit)

Estimation
In estimation, we round-off the numbers before carrying out the operations of addition,
subtraction, multiplication , division etc. For convenience, we will round-off each number
to the first non-zero digit, unless specified.

Approximate Numbers, Significant Digits, Scientific Notation,

A measurement consists of a numerical value and a unit of that measurement. Example: 4 kilograms,
where the 4 is the numerical value and the kilograms is the unit of the measurement.

Numbers obtained from a measurement are never exact (i.e., are approximate) due to the limitations
of the measuring instrument as well as the skill of the person making the measurement. As such,
when one records a measurement, one should indicate the reliability of the measurement. All
measurements may be assumed to have an uncertainty in at least one unit in the last digit of the
measurement, since in making a measurement, we usually estimate the last digit.

Results obtained from calculations using measurements are also as uncertain as the measurements
themselves.
In summary, the numbers that we deal with in calculations are obtained from observations.
 Some of the numbers are exact and some are approximate, The approximate numbers are
those numbers obtained from making measurements.

Exact Numbers: An exact number is a number that contains no uncertainties. It is assumed to be
infinitely accurate. We can obtain exact numbers from definitions and from direct count.
For example, the number of students in a math class by count is 25. In this case, there is no
uncertainty, since we know that there are exactly 25 students. Similarly, when one counts
200 dollars, one knows that one has exactly 200 dollars, and there is therefore, no uncertainty.
 Also by definition, 60 minutes = 1 hour; 2.54 centimeters = 1 inch. Since these numbers are
 defined, this 60 and 2,54 are exact and contain no uncertainties. We can also add that
this 60 has an infinite number of significant digits, (we can write 60 as 60.000...) and therefore
the zero in the 60 is significant. However if you make your own measurement and by coincidence
obtain 2.54 centimeters, then this 2.54 would not be exact. We can generalize that all the
conversion factors (from tables) are exact.

Significant Digits (Significant Figures), Digits obtained in a measurement

A significant digit (or figure) is one which is known to be reasonably reliable (or correct).
When we make measurements, the digits we read and estimate on a scale are also called
significant digits (or significant figures). These digits include digits that we are certain of,
 and one additional digit that we are uncertain of. This uncertain digit is obtained by the
estimation of the fractional part of the smallest subdivision on the scale being used.
As such, the rightmost digit is assumed to be uncertain.

Significant figure notation is an approximate method of indicating the uncertainty of
 a measurement. when recording a measurement.
We agree to the following:

1. The digits 1, 2, 3, 4, 5, 6, 7, 8, 9 are always significant.

2. The digit zero, 0. may or may not be significant according to its position in the number as
   follows:

  (a) Zeros before the first non-zero digits are **not** significant.

For example: (i) .0450 has **three** significant digits: The first zero is not significant; but the last zero is significant since if it were not we would not write it.

    (ii) .0012 has **two** significant digits. The first two zeros are not significant

.   (b) Zeros between non-zero digits **are** significant.

Example: 3.045 has **four** significant digits. The zero between 3 and 4 is significant.
    40.240 has **five** significant digits. The last zero is significant because if it were not we would not write it.

More examples: The numbers referred to below are assumed to have been obtained from measurements. 23.00 has four significant digits: the zeros in this case are significant since we do not have to write the zeros if they were not significant. If the zeros were not significant we would have written 23. 2300. has two significant digits, and 600 has one significant digit; however, in each of these two examples, the number of significant digits is sometimes ambiguous.

It is suggested that when the number of significant digits is in doubt , the maximum number of significant digits is to be assumed. Also, in recording data, if we know the number of significan digits, we will use the scientific notation.

Note above that if 600 and 2300 had been obtained by counting, or by definition, all the zeros would have been significant. As a reminder, significant notation generally pertains to numbers obtained from measurements.

Using **scientific notation** avoids all ambiguities with respect to the number of significant digits.. For instance, if we know that the above number, 600 were measured to two significant digits, then we would write 6.0×10^2. If the measurement were to three significant digits, we would write 6.00×10^2 When we deal with very large or very small numbers we prefer to write the numbers in scientific notation form. In this form, the scientific digits (digits) including the zeros can be unambiguously indicated. In scientific notation:

(1) 125000 would be written 1.25×10^5

(2) 1467 would be written 1.467×10^3

(3) .032500 would be written 3.2500×10^{-2}

(4) .0325 would be written 3.25×10^{-2}

(5) If 125000 were known to four significant digits it would be written 1.250×10^5.

Note: Some authors indicate which zeros are significant by underlining the last significant zero. For example, in 23000 the first two zeros are significant but the last zero is not significant. Note also that in some books, a decimal point placed after the last zero makes all the zeros significant. For example, 23000. has five significant digits, but 23000 has two significant digits. However, but there may still be ambiguity if the number is at the end of a sentence.

Accuracy and Precision in Measurements

Two contributions to uncertainty in measurement are limitations of precision, and limitations of accuracy. **Accuracy** indicates how close a measured value is to the true value but **precision** indicates how close two measurements of the same quantity are close to each other. Generally, more precision implies more accuracy. However, there are instances in which numbers may be more precise, but may not be more accurate.. For example, if a measuring device is incorrectly calibrated (having incorrect scale).

Accuracy and Precision in Calculations

With respect to significant digits, **accuracy** refers to the number of significant digits but **precision** refers to the number of decimal places. The larger the number of significant digits in a number, the more accurate the number. The larger the number of decimal places, the more precise the number.

Note: In calculations, the approximate numbers determine how the rounding-off is done

Rounding-off to Significant Digits or Figures in Arithmetic Operations
(Implicit Specification)

1. In **multiplication and division** involving significant digits, the product or quotient (answer) should be rounded-off so that the number of significant digits in the answer is equal to the number of significant digits of the number with the least number of significant digits. (In other words, the answer should not be more accurate than the number with the least accuracy)

Example 1: Multiply 2.34 cm by 5.6 cm

Step 1: $2.34 \times 5.6 = 13.104$

Step 2: The number with the fewest number of significant digits is 5.6 and it has two significant digits. Therefore, the product (answer) should contain only two significant digits.
Thus, 13.104 becomes 13.
Answer: 13 cm^2

In the above case the "4" in 2.34 is considered to be reasonably reliable. The " 6" in 5.6 is considered to be reasonably reliable.

Example 2: Maria determined the length of a piece of wood to be 6.47 yards. What is the length of this wood in feet?
Solution
By definition, 3 feet = 1 yard
6.47 yards is used to determine the number of significant digits in the answer,
(The 3 feet is exact and has an infinite number of significant digits)

$$\frac{6.47 \text{ yards}}{1} \times \frac{3 \text{ feet}}{1 \text{ yard}}$$

= 19.41 feet

= 19.4 feet (6.47 has three significant figures)

2. In addition and subtraction, we shall round-off so that the number of decimal place in the answer equals the number of decimal places in the number with the least number of decimal places. (In other words, the answer should not be more precise than the number with the least precision)

Example 1: Add: 143.54, 172.3, and 64.62

Solution

143.54 < - - - - has two decimal places
172.3 < - - - - -has one decimal place (determines the number of decimal places in answer)
 64.62 < - - - - - has two decimal places
380.46
Answer: 380.5 (has one decimal place)

Example 2 Subtract 12.4 from 143.63
Solution

143.63 < - - - - has two decimal places

 12.4 < - - - - -has one decimal place (determines the number of decimal places in answer)

131.23
Answer: 131.2 (has one decimal place)

3. **In finding powers and roots,** the root or power should be rounded-off so that the number of significant digits in the answer is equal to the number of significant digits in the number.

Example 1: Find $\sqrt{26.9}$
Radicand has three significant digits , and therefore the root should have three significant digits.
From a calculator, $\sqrt{26.9} = 5.1865$

  $\sqrt{26.9} = 5.19$ (has three significant digits. Same as in the radicand)

Note: When two or more different operations are involved, the final operation determines how the final result is rounded-off.
Example Simplify: $38.3 + 12.9(3.58)$
Solution According to order of operations, we multiply 12.9 by 3.58 first and then add 38.3

  $38.3 + 12.9(3.58)$

$= 38.3 + 46.182$

$= 84.485$

$= 84.5$ (has one decimal place as in 38.3)

Since addition was the last step, we use precision (the number of decimal places) to round-off.

Addition and Subtraction Involving Scientific Notation   173

Before adding or subtracting the numbers must have the **same powers** of 10. We will rewrite the expression so that the power of 10 is that of the highest power in the expression

Example: **1.** $4 \times 10^2 + 3 \times 10^2 = (4 + 3) \times 10^2$
$$= 7 \times 10^2$$

Example: **2.** $4 \times 10^2 + 2 \times 10^3 = 0.4 \times 10^3 + 2 \times 10^3$
$$= (0.4 + 2) \times 10^3$$
$$= 2.4 \times 10^3$$

or

$4 \times 10^2 + 2 \times 10^3 = 4 \times 10^2 + 20 \times 10^2$
$$= (4 + 20) \times 10^2$$
$$= 24 \times 10^2$$
$$= 2.4 \times 10^3 \quad \text{(Again, we obtain the same result)}$$

Order of Magnitude (for comparing relative sizes using powers of 10).
The order of magnitude is the power of 10 closest to the given number.
(It is an approximation to the number. Note the sequence, $..., 10^{-2}, 10^{-1}, 10^0, 10^1, 10^2, ...$)

If a given quantity is 1000 times another quantity, the given quantity is larger by three orders of magnitude.
Examples:
1. The order of magnitude of 123 is 10^2, since 123 is closer to 100 than to 1000.

2. Find the order of magnitude of 0.00352.
Solution
Step 1: Write the number in scientific notation.
$$0.00352 = 3.52 \times 10^{-3}.$$
Step 2: Since the integer before the decimal point, 3, is less than 5, we replace 3.52 by 1
(since this is closer to 1 or 10^0, than it is to 10 or 10^1)
Step 3: 3.52×10^{-3}
$$= 10^0 \times 10^{-3}$$
$$= 1 \times 10^{-3}$$
$$= 10^{-3}$$
The order of magnitude of 0.00352 is 10^{-3}. (since by definition, the order of magnitude is the power of 10 closest to the given number.).
We can use the order of magnitude in estimation by rounding-off to the orders of magnitude.

More examples: Round-off to the nearest order of magnitude.

1. 1.32×10^2

2. 8.02×10^4

3. 0.0009

4. 0.0302

Solution:

1. Since 1.32 is closer to 1 than to 10,

$$1.32 \times 10^2$$
$$= 10^0 \times 10^2$$
$$= 1 \times 10^2$$
$$= 10^2$$

The order of magnitude of 1.32×10^2 is 10^2

2. 8.02 is closer to 10 than to 1

$$8.02 \times 10^4$$
$$= 10^1 \times 10^4$$
$$= 10^5$$

The order of magnitude is 10^5

3. $\quad 0.0009 = 9 \times 10^{-4}$
$$= 10^1 \times 10^{-4}$$
$$= 10^{-3}$$

.

The order of magnitude is 10^{-3}.

4. $0.0201 = 2.01 \times 10^{-2}$
$$= 10^0 \times 10^{-2}$$
$$= 1 \times 10^{-2}$$
$$= 10^{-2}$$

.

The order of magnitude is 10^{-2}

Summary for rounding-off a number to the order of magnitude.

Step !: Write the number in scientific notation

Step 2: Ignoring the power of 10, if the integer before the decimal point is 5 or greater, replace the non-power of 10 part by 10 ((i.e. 10^1); but if the integer is less than 5, replace the non-power of 10 part by 1 (10^0)

Step 3: Simplify

International System of Units

The International System of Units (SI) has adopted a set of seven base (or primary) units.

Quantity	Unit	Symbol
Length	meter	m
Mass	kilogram	kg
Time	seconds	s
Electric current	ampere	A
Temperature	Kelvin	K
Amount of substance	mole	mol
Luminous Intensity	candela	cd

Derived units

In addition to the seven base units, there are derived units which are combinations of the base units

Example:

From the SI base unit, m (meter). for length, the unit for area is $m \times m = m^2$,
Since area = length × width, and the unit of length is m and the unit of with is m.

For more practical or convenient units, we use prefixes and multiplication factors (in powers of 10) to express other units.

Example: 1 kilometer = $10^3 m$, where kilo = 10^3

$1 \text{ km} = 10^3 \text{ m}$ or I km = 1000m.

INDEX

E

F

G

I

L

M

INDEX

Some useful conversion factors

Units of length
1 cm = .3937 in = .0328ft = .01094 yd
1 m = 100 cm = 39.3701 in = 3.2808 ft.= 1.0936 yd

1 in. = 2.54 cm = .0833 ft = .0254 m
1 ft = 12 in. = 30.48 cm = .3048 m = .3333 yd
1 mile = 1760 yd = 5280 ft = 1.6093 km
1 yd = 3 ft = 36 in.= .9144 m = 91.44 cm
1 km = .62137 mile = 1000 m = 100,000 cm

Units of mass
1 kg = 1000 g = 2.2046 lb =35.274 oz
1 lb (avdp) = 453.592 g = 16 oz

1 metric ton = 1000 kg = 10^6 g = 1.1023 ton
1 ton = 907.1847 kg = 2000 lb = .9072 metric ton
1 gm = .03527 oz = 1000 mg =.0022046 lb
1 oz = 28.3495 g = .0625 lb = 16 drams
1 long ton (British) = 2240 lb

Units of volume
1 liter = 1000 cm^3 = 61.0237 $in.^3$= .26417 gal = 1.0567 qt =.03531 ft^3= 2.113 pt

1 gal (U.S.) = 4 qt = 3.7854 liter = 8 pt = 231 $in.^3$ = .13368 ft^3

1 qt = 2 pt = .946353 liter =946.353 cm^3 = 57.75 $in.^3$ = .25 gal =.034201 ft^3

1 cord = 128 ft^3
1 pt = .473 liter = .5 qt
1 ft^3 =1728 $in.^3$
1 yd^3 = 27 ft^3 = 46656 $in.^3$

Units of area
1 ft^2 (sq. ft) = 144 $in.^2$ (sq. in.)

1 yd^2 (sq. yd) = 9 ft^2 (sq. ft) = 1296 sq. in.

1 $mile^2$ (sq. mile) = 640 acres

1 m^2 = $10^4 cm^2$

1 acre = 4840 yd^2

Units of area
1 ft^2 (sq. ft) = 144 $in.^2$ (sq. in.)

1 yd^2 (sq. yd) = 9 ft^2 (sq. ft) = 1296 sq. in.

1 $mile^2$ (sq. mile) = 640 acres

1 m^2 = $10^4 cm^2$

1 acre = 4840 yd^2

Symbols for units

cm = centimeter	gal = gallon	g = gram
m = meter	qt = quart	kg = kilogram
in. = inch	pt = pint	lb = pound
ft = foot	oz = ounce	mg = milligram
yd = yard		
km = kilometer		
mi = mile		

Some prefixes (International System)

Prefix	Power
tera	10^{12}
giga	10^9
mega	10^6
kilo	10^3
hecto	10^2
deka	10^1
deci	10^{-1}
centi	10^{-2}
milli	10^{-3}
micro	10^{-6}
nano	10^{-9}
pico	10^{-12}

Conversion Factors for Measurements

American System (British System) Interconversion (Factors) Metric System

Some " **bridge**s" for converting from one system to the other

Length

12 inches (in) = 1 foot (ft.)
3 feet (ft.) = 1 yard (yd)
5280 feet = 1 mile (mi)
1760 yards = 1 mile

1 in. = 2.54 cm
1 yd = 0.9144 m
1 mi = 1.61 km
1 km = 0.62 mi

1 kilometer (km) $= 10^3$ m = 1000 m
1 hectometer (hm) $= 10^2$ m = 100 m
1 dekameter (dam) $= 10^1$m = 10 m
1 meter (m) $= 10^0$m = 1m
1 decimeter (dm) $= 10^{-1}$ m = 0.1m
1 centimeter (cm) $= 10^{-2}$ m = 0.01m
1 millimeter (mm) $= 10^{-3}$ m = 0.001m

Some " **bridge**s" for converting from one system to the other

Mass

1 lb = 16 oz
1 ton = 2000 lb
1 long ton = 2240 lb

1 kg = 2.2 lb
1 lb = 454 g
1 oz = 28.4 g = 16 drams
1 ton = 0.9072 metric ton

1 kilogram (kg) $= 10^3$ g = 1000 g
1 hectogram (hg) $= 10^2$ g = 100 g
1 dekagram (dag) = 10^1 g = 10 g
1 gram (g) $= 10^0$ g = 1 g
1 decigram (dg) $= 10^{-1}$ g = 0.1g
1 centigram (cg)) $= 10^{-2}$ g = 0.01g
1 milligram (mg) = 10^{-3} g = 0.001 g

Some " **bridge**s" for converting from one system to the other

Volume

16 fluid oz (fl-oz)= 1 pint (pt)
2 pints (pt) = 1 quart (qt)
4 quarts = 1 gallon (gal)

1 liter (l) = 1.057 qt
1 gal = 3.785 l
1 liter = 2.1 pt
1 pt = .473 l

1 kiloliter (kl) $= 10^3$ l = 1000 l
1 hectoliter (hl) $= 10^2$ l = 100 l
1 dekaliter (dal) = 10^1 l = 10 l
1 liter (l) $= 10^0$l = 1 l
1 deciliter (dl) $= 10^{-1}$l = 0.1l
1 centiliter (cl) = 10^{-2}l = 0.01l
1 milliliter (ml) = 10^{-3}l = 0.001 l

Mnemonic device (metric system)

Must remember the following (metric system:)

100 cm = 1 m
1000 m = 1 km

1000 mg = 1 g
1000 g = 1 kg

1000 ml = 1 l
1 ml = 1 cc = 1cm^3
1000 cc = 1 l

k – ilo – 10^3
h – ecto – 10^2
d – eka – 10^1
d – ec*i* – 10^{-1}
c – enti – 10^{-2}
m – illi– 10^{-3}

Say the following aloud:
Step 1: First go down vertically as kei-eitch-dii-dii-see-em, then Step 2
Step 2: Kilo-hecto-deka-deci-centi-milli, and then note how the powers
        decrease vertically downwards.
    Examples: 1 Kilometer = 10^3 meter; 1 milligram =10^{-3}gram;
    1 centimeter =10^{-2} meter =$\frac{1}{100}$ meter --->100 centimeters = 1 meter.

Mathematical Modeling
Some Reciprocal Relationships

1. Arithmetic If A working alone can do a piece of work in time t_A; B working alone can do the same work in time t_B; C working alone can do the same work in time t_C, and if A, B, and C working together, can do the same work in time t_{ABC}, then

$$\frac{1}{t_{ABC}} = \frac{1}{t_A} + \frac{1}{t_B} + \frac{1}{t_C}$$

That is, the reciprocal of the working-together time equals the sum of the reciprocals of working-alone times (individual times).

2. Geometry: For any triangle, the reciprocal of the inradius (R) equals the sum of the reciprocals of the exradii (r_1, r_2, and r_3).

Thus
$$\frac{1}{R} = \frac{1}{r_1} + \frac{1}{r_2} + \frac{1}{r_3}$$

3. Physics (Electricity) For electrical resistances in parallel (in an electric circuit), the reciprocal of the combined resistance, R, equals the sum of the reciprocals of the separate resistances, r_1, r_2, and r_3.

Thus
$$\frac{1}{R} = \frac{1}{r_1} + \frac{1}{r_2} + \frac{1}{r_3}$$

4. Physics (Optics)

For two thin lenses in contact, the reciprocal of the combined focal length, F, equals the sum of the reciprocals of the separate focal lengths, f_1 and f_2, .

Thus
$$\frac{1}{F} = \frac{1}{f_1} + \frac{1}{f_2}$$

5. Physics (Optics) For spherical mirrors and thin lenses, the reciprocal of the focal length F equals the sum of the reciprocals of the object distance, d_o and the image distance d_i.

Thus
$$\frac{1}{F} = \frac{1}{d_o} + \frac{1}{d_i}$$

6. Physics (Mechanics). If two bubbles of radii r_1, r_2, coalesce into a double bubble, the radius, R, of the partition is given by

$$\frac{1}{R} = \frac{1}{r_1} - \frac{1}{r_2}$$